Cambridge Tracts in Mathematics and Mathematical Physics

No. 53

TOPOLOGICAL VECTOR SPACES

TOPOLOGICAL VECTOR SPACES

BY

A. P. ROBERTSON

Professor of Mathematics
Murdoch University, Western Australia

AND

WENDY ROBERTSON

Senior Lecturer in Mathematics
University of Western Australia

SECOND EDITION

CAMBRIDGE
AT THE UNIVERSITY PRESS
1973

Published by the Syndics of the Cambridge University Press
Bentley House, 200 Euston Road, London NW1 2DB
American Branch: 32 East 57th Street, New York, N.Y.10022

Library of Congress Catalogue Card Number: 72–89805

ISBN: 0 521 20124 1

First published 1964
Reprinted 1966
Second edition 1973

Printed in Great Britain
at the University Printing House, Cambridge
(Brooke Crutchley, University Printer)

CONTENTS

PREFACE

The aim of this tract is to give an account of the theory of topological vector spaces. There are many branches of functional analysis which depend upon this theory; it forms, for example, the background for the theory of distributions. In addition, it often clarifies results in the theory of normed spaces, especially those concerned with the weak topology, to regard them as particular cases of more general results about topological vector spaces. In this tract we have tried to present the main theorems as simply and directly as is possible without sacrificing any of their force or generality. It has been our purpose to make the basic ideas and facts of the subject easily accessible. With this in mind, we have assumed only a minimum of knowledge of general topology and linear algebra to start with; the details of what is assumed are to be found at the beginning of the first chapter. After this, new topological concepts are introduced when they are needed. The first four or five chapters contain the fundamental results; here the pace is leisurely. Some of these could have been deduced more concisely from general topological theorems, but we have preferred to keep the treatment self-contained and establish them directly for topological vector spaces, or even for locally convex spaces. From the sixth chapter onwards the proofs are more condensed; here we have allowed ourselves to be guided to some extent by our own interests in the choice of subject matter.

Each of the first six chapters has a supplementary section containing illustrative examples and outlining further developments of topics discussed in the chapter. Although the text is self-contained, we have felt free in the supplements to use mathematical notions not, or not yet, defined in the text. Where there are alternative notations and terms in current use, we have, for the most part, followed those in the treatise of Bourbaki. The numbering of theorems, propositions and lemmas is started afresh at the beginning of each chapter; a reference not preceded by a chapter number applies to the chapter in which it occurs.

We have made no attempt to ascribe each result to its original sources, except that the well-known theorems are given their customary names.

It is a pleasure to both of us to record our gratitude to Dr F. Smithies. Not only did he read our typescript and give us valuable advice and help at all stages of preparation of this tract, but it was he who first aroused our interest in functional analysis by his lectures and seminars. We should also like to thank the Editors of the Cambridge Mathematical Tracts for help with proof correction and the Cambridge University Press for their careful printing. During the first year of preparation of this tract we held research fellowships from St John's College, Cambridge, and Girton College, Cambridge, and we are glad now to have the opportunity of acknowledging this assistance.

GLASGOW
March 1964

A.P.R.
W.J.R.

PREFACE TO THE SECOND EDITION

In this edition, the section on the closed graph theorem has been brought up to date by the inclusion of an account of spaces with webs. In order not to disturb the original arrangement of the book, we have placed this new material in an Appendix. The only other change is the removal of a number of errors and obscurities; we are grateful to readers who have pointed them out. We are glad to thank also the staff of the Cambridge University Press for their unfailing helpfulness and courtesy.

KEELE
July 1972

A. P. R.
W. J. R.

CHAPTER I

DEFINITIONS AND ELEMENTARY PROPERTIES

A topological vector space is a set with two compatible structures. On the one hand, it has the algebraic structure of a vector space and on the other, it has a topology, so that the notions of convergence and continuity are meaningful. These two structures must be compatible in the sense that the algebraic operations are continuous. In the first two sections of this chapter, we review briefly the definitions and some of the most elementary properties of vector spaces and topological spaces; one or two results on convex sets also appear in § 1. The formal development of the theory begins in the third section, and the remainder of this chapter contains an account of some of those main properties of topological vector spaces that are direct consequences of the definition.

1. Vector spaces. A vector space over the field of real or complex numbers is a natural generalisation of ordinary three-dimensional Euclidean space. There are two algebraic operations, the addition of vectors and the multiplication of a vector by a scalar, and they have to satisfy some reasonable conditions. Precisely, a vector space E over the field Φ of real numbers is a set admitting addition (any two elements x, y of E possess a sum $x+y$ which is an element of E) and multiplication by scalars (for each x in E and λ in Φ, the product λx is defined and is an element of E) with the following properties:

(1) $x+y = y+x$;

(2) $x+(y+z) = (x+y)+z$;

(3) there is a zero element, or origin, in E, denoted by o, such that for all x in E, $o+x = x$;

(4) for each x in E there is an element $-x$ such that

$$x+(-x) = o$$

(and we write $x-y$ for $x+(-y)$);

(5) $(\lambda\mu)x = \lambda(\mu x)$;
(6) $(\lambda+\mu)x = \lambda x+\mu x$;
(7) $\lambda(x+y) = \lambda x+\lambda y$;
(8) $1.x = x$.

From these, it follows that the zero is unique and that, for each x in E, the element $-x$ is unique; it is easy to show that $0.x = o$ and $(-1).x = -x$ for all x in E and $\lambda.o = o$ for all λ in Φ. (In fact (1)–(8) form a sufficient set of axioms.) A vector space over the field of complex numbers can be described in exactly the same way with the word 'complex' substituted for the word 'real'. The contents of this tract are for the most part equally applicable whether the scalar field is real or complex; wherever possible, the results and proofs have been arranged so as to make it unnecessary to specify one or other scalar field. The main exceptions are the proofs of the Hahn–Banach theorem (Ch. II, Th. 2) and the Krein–Milman theorem (Ch. VII, Th. 1) where we deal separately with the real and complex cases.

We shall use $\{x: P(x)\}$ for the set of elements x possessing the property $P(x)$. It is useful to extend to subsets of E the notation for linear combinations of elements:

if $x \in E$ and $A \subseteq E$, $x+A = \{x+y: y \in A\}$;
if $A \subseteq E$ and $B \subseteq E$, $A+B = \{x+y: x \in A, y \in B\}$;
if $\lambda \in \Phi$ and $A \subseteq E$, $\lambda A = \{\lambda x: x \in A\}$.

Also if $\mathscr{A}$ is a set of subsets of E, $x+\mathscr{A} = \{x+A: A \in \mathscr{A}\}$.

Unfortunately, the set $A+A$ is in general distinct from the set $2A$; however, it is always true that $2A \subseteq A+A$.

A subset M of E is called a *vector subspace* of E if $x+y \in M$ and $\lambda x \in M$ for all $x \in M$, $y \in M$ and $\lambda \in \Phi$ (i.e. if $M+M \subseteq M$ and $\lambda M \subseteq M$ for all $\lambda \in \Phi$). The set $\{o\}$ consisting of the origin alone is a vector subspace. Any intersection of vector subspaces is a vector subspace. Given any subset A of E, the set of all finite linear combinations $\sum_i \lambda_i x_i$ with $\lambda_i \in \Phi$ and $x_i \in A$ is a vector subspace of E, called the *vector subspace spanned by* A (or the vector subspace generated by A). It is the intersection of all the vector subspaces of E containing A.

The subset A of E is called *linearly independent* if

$$\sum_{1 \leqslant i \leqslant n} \lambda_i x_i = o$$

implies $\lambda_1 = \lambda_2 = \ldots = \lambda_n = 0$ for every choice of n and of $x_1, x_2, \ldots, x_n$ in A. A linearly independent subset spanning E is called a *base* of E. Clearly any *maximal* linearly independent set (i.e. one to which no new element can be adjoined without destroying linear independence) is a base of E, and any *minimal* set spanning E is also a base of E.

If a linearly independent set is not a base we can add elements to it one at a time, producing a succession of linearly independent subsets that form a chain: a set $\mathscr{C}$ of subsets of E is called a *chain* if, of any two sets of $\mathscr{C}$, one is contained in the other. It is not at all obvious that by continuing in this way we could ever reach a base of E; to show that a base always exists it is necessary to invoke some form of the axiom of choice. Here and later when we need an application of the axiom of choice we shall find it convenient to use it in the following form.

MAXIMAL AXIOM. *Let $\mathscr{E}$ be a set of subsets of a set E and let $\mathscr{C}$ be a chain contained in $\mathscr{E}$. Then there is a maximal chain $\mathscr{M}$ with $\mathscr{C} \subseteq \mathscr{M} \subseteq \mathscr{E}$.*

This axiom is equivalent to Zorn's lemma and so also to the axiom of choice.† In fact, in our uses of the axiom, we shall find that we apply it only in cases when $\mathscr{C}$ is a trivial chain consisting of one set C alone. The axiom then asserts that a subset C of E belonging to a set $\mathscr{E}$ of subsets of E can be embedded in a maximal chain $\mathscr{M} \subseteq \mathscr{E}$.

THEOREM 1. *If L is a linearly independent subset of a vector space E and if S is a subset containing L and spanning E, there is a base B of E with $L \subseteq B \subseteq S$.*

Proof. Let $\mathscr{E}$ be the set of all linearly independent subsets of S and $\mathscr{C} = \{L\}$. By the axiom there is a maximal chain $\mathscr{M}$ with $\mathscr{C} \subseteq \mathscr{M} \subseteq \mathscr{E}$. Let B be the union of the sets of $\mathscr{M}$. Then $L \subseteq B \subseteq S$. Also B is linearly independent, for any finite linear combination of elements of B is contained in a finite union of sets of $\mathscr{M}$ and so in the largest of them, $\mathscr{M}$ being a chain. Finally, every element of S is a linear combination of elements of B, because if one,

† There is a full treatment of this topic in Chapter O of J. L. Kelley's book, *General topology* (Van Nostrand, New York, 1955).

x say, were not, $B \cup \{x\}$ could be added to $\mathscr{M}$, and this would contradict the maximality of $\mathscr{M}$.

COROLLARY. *Every vector space has a base.* (Take $L = \emptyset$, the empty set, and $S = E$.)

If a vector space E has a base containing a finite number n of elements, then every base has n elements and every set of n elements that either is linearly independent or spans E is a base; E is then called *n-dimensional*. The scalar field itself is a vector space of dimension 1 over itself; any non-zero member of Φ forms a base.

A subset A of a vector space E is called *convex* if, for all $x \in A$ and $y \in A$, $\lambda x + \mu y \in A$ whenever $\lambda \geqslant 0, \mu \geqslant 0$ and $\lambda + \mu = 1$. It is called *balanced* if, for all $x \in A$, $\lambda x \in A$ whenever $|\lambda| \leqslant 1$. The set A is called *absolutely convex* if it is both convex and balanced; this is equivalent to saying that, for all $x \in A$ and $y \in A$, $\lambda x + \mu y \in A$ whenever $|\lambda| + |\mu| \leqslant 1$. For clearly a set with this property is convex and balanced; conversely, suppose that A is convex and balanced, that $x \in A$, $y \in A$ and $|\lambda| + |\mu| \leqslant 1$. If $\lambda = 0$ or $\mu = 0$, then certainly $\lambda x + \mu y \in A$. If $\lambda \neq 0$ and $\mu \neq 0$, then $(\lambda/|\lambda|)\,x \in A$, $(\mu/|\mu|)\,y \in A$ and $|\lambda|/(|\lambda| + |\mu|) + |\mu|/(|\lambda| + |\mu|) = 1$, so that

$$\lambda x + \mu y = (|\lambda| + |\mu|)\left(\frac{|\lambda|}{|\lambda| + |\mu|}\,\frac{\lambda x}{|\lambda|} + \frac{|\mu|}{|\lambda| + |\mu|}\,\frac{\mu y}{|\mu|}\right) \in A.$$

It follows immediately from the definitions that if A is convex, so is $x + \lambda A$ for each $x \in E$ and $\lambda \in \Phi$; if A and B are absolutely convex, then so are $A + B$ and λA, for each $\lambda \in \Phi$.

Some useful properties of absolute convexity are summarised, for reference, in

LEMMA 1. *Suppose that A is a non-empty absolutely convex set. Then*

(i) $o \in A$,

(ii) $\lambda A \subseteq \mu A$ *whenever* $|\lambda| \leqslant |\mu|$, *and*

(iii) $\sum_{1 \leqslant i \leqslant n} (\lambda_i A) = (\sum_{1 \leqslant i \leqslant n} |\lambda_i|)\,A$

for all $\lambda_i \in \Phi$.

Proof. (i) Let $x \in A$; then $o = \frac{1}{2}x + (-\frac{1}{2})\,x \in A$. (ii) If $\mu = 0$, then $\lambda A = \mu A = \{o\}$; if $\mu \neq 0$ and $x \in A$ then $|(\lambda/\mu)| \leqslant 1$ and so $(\lambda/\mu)\,x \in A$ since A is balanced. Hence $\lambda x \in \mu A$; thus $\lambda A \subseteq \mu A$.

(iii) For the case $n = 2$ we have to prove $\lambda A + \mu A = (|\lambda| + |\mu|) A$. If $\lambda = \mu = 0$ this is trivial. If $\lambda \neq 0$ or $\mu \neq 0$ then

$$\frac{\lambda}{|\lambda| + |\mu|} A + \frac{\mu}{|\lambda| + |\mu|} A \subseteq A$$

and so $\lambda A + \mu A \subseteq (|\lambda| + |\mu|) A$. On the other hand,

$$(|\lambda| + |\mu|) A \subseteq |\lambda| A + |\mu| A \subseteq \lambda A + \mu A$$

by (ii). The general case follows by induction on n.

Any intersection of convex sets is convex. Given any subset A of E, the set of all finite linear combinations $\sum_i \lambda_i x_i$ with $\lambda_i \geqslant 0$, $\sum_i \lambda_i = 1$ and each $x_i \in A$ is a convex set containing A called the *convex envelope* of A. It is the intersection of all the convex subsets of E containing A and so is the smallest such subset. The *absolutely convex envelope* of A is the set of all finite linear combinations $\sum_i \lambda_i x_i$ with $\sum_i |\lambda_i| \leqslant 1$ and each $x_i \in A$, and is the smallest absolutely convex set containing A.

Suppose that A is an absolutely convex set spanning E. Then, if $x \in E$, there is some $\lambda > 0$ with $x \in \lambda A$. For then $x = \sum_i \lambda_i x_i$ with each $x_i \in A$ and, by Lemma 1 (iii), $\sum_i \lambda_i x_i \in (\sum_i |\lambda_i|) A$, so that $\lambda = \sum_i |\lambda_i|$ will suffice. Moreover, by Lemma 1 (ii), $x \in \mu A$ for all μ with $|\mu| \geqslant \lambda$.

Any subset A of a vector space E is called *absorbent* if for each $x \in E$ there is some $\lambda > 0$ such that $x \in \mu A$ for all μ with $|\mu| \geqslant \lambda$. A finite intersection of absorbent sets is clearly absorbent. An absolutely convex set is absorbent if and only if it spans E; this is equivalent to

$$E = \bigcup_{\lambda > 0} \lambda A,$$

or, by Lemma 1 (ii),

$$E = \bigcup_{n=1}^{\infty} nA.$$

2. Topological spaces. A topological space is a set provided with a structure that enables convergence and continuity to be considered. One method of imposing such a structure is to specify which subsets are to be called *open*. Guided by the familiar cases of the real numbers and the plane, we demand that the following axioms be verified:

O 1: *every union of open sets is open;*

O 2: *every finite intersection of open sets is open;*

O 3: *the empty set and the whole space are open.*

A set E in which is defined a set of open subsets satisfying these three axioms is called a *topological space* and its elements are often called points.

Given a point x of a topological space E, the set U is called a *neighbourhood* of x if there is an open set V with $x \in V \subseteq U$.

The point x is called an *interior point* of the subset A of E if there is a neighbourhood U of x contained in A. The set of all interior points of A is an open set contained in A, called the *interior* of A, and A is open if and only if it is identical with its interior.

A set A is called *closed* if its complement $\sim A$ is open. Any intersection, or any finite union, of closed sets is closed; $\emptyset$ and E are both closed. The point x is called a *point of closure* of the set A if every neighbourhood of x meets A. The set of all points of closure of A is a closed set containing A called the *closure* of A and denoted by $\bar{A}$. The set A is closed if and only if it is identical with its closure.

If A and B are subsets of E, A is called *dense* in B if $B \subseteq \bar{A}$.

If $\mathscr{U}_x$ denotes the set of all neighbourhoods of the point x of the topological space E, then $\mathscr{U}_x$ has the following properties:

N 1: $x \in U$ *for all* $U \in \mathscr{U}_x$;

N 2: *if* $U \in \mathscr{U}_x$ *and* $V \in \mathscr{U}_x$ *then* $U \cap V \in \mathscr{U}_x$;

N 3: *if* $U \in \mathscr{U}_x$ *and* $U \subseteq V$ *then* $V \in \mathscr{U}_x$;

N 4: *if* $U \in \mathscr{U}_x$, *there is some* $V \in \mathscr{U}_x$ *with* $U \in \mathscr{U}_y$ *for all* $y \in V$.

(In N 4, the interior of U is a suitable V.)

Conversely, suppose that E is any set, and that for each element x of E is given a (non-empty) set $\mathscr{U}_x$ of subsets of E. Then, if the conditions N 1–N 4 are satisfied, exactly one open-set topology can be defined on E in such a way as to make $\mathscr{U}_x$ the set of neighbourhoods of x for each x. (Call the set $A \subseteq E$ open if for each $x \in A$, there is some $U \in \mathscr{U}_x$ with $U \subseteq A$.)

A subset $\mathscr{V}_x$ of the set $\mathscr{U}_x$ of neighbourhoods of x is called a *base* of neighbourhoods of x if, given $U \in \mathscr{U}_x$, there is a $V \in \mathscr{V}_x$ with $V \subseteq U$. The open sets containing x form such a base. If $\mathscr{V}_x$ is a base of neighbourhoods of x then $\mathscr{U}_x$ is the set of all $U \subseteq E$ such that there exists a $V \in \mathscr{V}_x$ with $V \subseteq U$.

It is possible to define different topologies on the same set E. If ξ and η are two topologies on E we say that ξ is *finer* than η (or η is *coarser* than ξ) if every set that is open in η is also open in ξ. If $\mathscr{U}_x^\xi$ is the set of neighbourhoods of x in the topology ξ and $\mathscr{V}_x^\xi$ is a base of these neighbourhoods then ξ is finer than η if and only if $\mathscr{U}_x^\eta \subseteq \mathscr{U}_x^\xi$ for all x in E, or equivalently if, given $U \in \mathscr{V}_x^\eta$, there is a $V \in \mathscr{V}_x^\xi$ with $V \subseteq U$.

Our use of 'finer' allows the two topologies to be identical. Two topologies on E may fail to be comparable; each may have open sets that are not open for the other.

A topological space is called *separated* (or *Hausdorff*) if any two distinct points have disjoint neighbourhoods. If E is separated under the topology ξ, then E is separated under any topology finer than ξ.

Metric spaces form an important class of topological spaces. A real-valued function d, defined for each pair of elements x, y of a set E is called a *metric* if it satisfies

M 1: $d(x,y) \geqslant 0$, $d(x,x) = 0$ and $d(x,y) > 0$ if $x \neq y$;

M 2: $d(x,y) = d(y,x)$;

M 3: $d(x,z) \leqslant d(x,y) + d(y,z)$, the triangle inequality.

A set E provided with a metric is called a *metric space* and $d(x,y)$ is called the *distance* between x and y. Let $V(x,\epsilon)$ be the set of all elements y of E such that $d(x,y) < \epsilon$. Then, if we call U a neighbourhood of x if $V(x,\epsilon) \subseteq U$ for some $\epsilon > 0$, the set $\mathscr{U}_x$ of all neighbourhoods of x satisfies N 1–N 4 and so defines a topology on E. A topological space is called *metrisable* if its topology can be defined by a metric d. Different metrics may define the same topology; they are then called *equivalent*. As an example of this, suppose that d is a metric on E; then $d' = \inf\{1, d\}$ is an equivalent metric for which $d'(x,y) \leqslant 1$ for all $x, y \in E$. A metrisable space is separated, because if $x \neq y$, $d(x,y) = \alpha > 0$ and so $V(x, \frac{1}{2}\alpha)$ and $V(y, \frac{1}{2}\alpha)$ are disjoint neighbourhoods of x and y. Every point x of a metrisable space has a countable base of neighbourhoods, e.g. the base $\{V(x, 1/n): n = 1, 2, \ldots\}$.

The real numbers and the complex numbers can both be topologised by taking $d(x, y) = |x-y|$. This is their natural topology and the only one we shall consider on these two sets. On n-dimensional space, with points $x = (x_1, x_2, \ldots, x_n)$, $y = (y_1, y_2, \ldots, y_n)$, the distances

$$\max_{1 \leqslant i \leqslant n} |x_i - y_i|, \qquad \sqrt{\sum_{1 \leqslant i \leqslant n} |x_i - y_i|^2}, \qquad \sum_{1 \leqslant i \leqslant n} |x_i - y_i|$$

are equivalent metrics.

If E is a topological space and H a subset of E, H can be made into a topological space by taking the intersections with H of the open sets of E to be the open sets of H. The axioms O 1, O 2, and O 3 are easily verified and the topology so defined on H is called the *induced topology*. The closed subsets of H in this topology are the intersections with H of the closed subsets of E, and the neighbourhoods of any point of H are the intersections with H of the neighbourhoods of this point in E. If E is separated or metrisable, H has the same property.

Let E and F be two topological spaces. The function f mapping E into F is called *continuous at the point* x of E if, to each neighbourhood V of $f(x)$ in F corresponds a neighbourhood U of x in E such that if $x \in U$, then $f(x) \in V$, i.e. $f(U) \subseteq V$. If $f^{-1}(B)$ denotes the set of all x in E with $f(x) \in B$, f is continuous at x if and only if $f^{-1}(V)$ is a neighbourhood of x for every neighbourhood V of $f(x)$. If f is continuous at each point of E, f is said to be *continuous* (on E). The following two conditions are each equivalent to the continuity of f: $f^{-1}(B)$ is an open subset of E for every open subset B of F; $f^{-1}(B)$ is a closed subset of E for every closed subset B of F.

A (1, 1) mapping f of E onto F is called *bicontinuous* if both f and its inverse f^{-1} are continuous. Then f puts E and F, together with their topologies, into (1, 1) correspondence; we say that E and F are *homeomorphic*, and that f is a *homeomorphism*.

Let E, F and G be three topological spaces. Let f be a continuous mapping of E into F and g a continuous mapping of F into G. If we put $h(x) = g(f(x))$, h is a continuous mapping of E into G, denoted by $g \circ f$. If f and g are homeomorphisms, so is $g \circ f$, and $(g \circ f)^{-1} = f^{-1} \circ g^{-1}$.

3. Topological vector spaces. Let E be a vector space over the real or complex field Φ. A topology ξ on E is said to be *compatible* with the algebraic structure of E if the algebraic operations in E are continuous, i.e. $x+y$ is a continuous function of the pair of variables x, y, and λx is a continuous function of the pair of variables λ, x. A *topological vector space* over Φ is a vector space over Φ with a compatible topology.

The following propositions give the consequences of the continuity conditions.

PROPOSITION 1. *For each $a \in E$ the translation $f: f(x) = x+a$ is a homeomorphism of E onto itself. In particular, if $\mathcal{U}$ is a base of neighbourhoods of the origin, $\mathcal{U}+a$ is a base of neighbourhoods of a.*

Proof. If $f(x) = x+a = y$, then $f^{-1}(y) = x = y-a$. Thus f is a $(1, 1)$ mapping of E onto itself, which, together with its inverse, is continuous. Hence f is a homeomorphism.

Thus the whole topological structure of E is determined by a base of neighbourhoods of the origin. We therefore work mainly with neighbourhoods of the origin, and where it will cause no confusion, we call them simply 'neighbourhoods'. If U is a neighbourhood (of the origin), $U+a$ is the corresponding neighbourhood of a, and $x \in U+a$ if and only if $x-a \in U$.

PROPOSITION 2. *For each non-zero $\alpha \in \Phi$ the mapping f: $f(x) = \alpha x$ is a homeomorphism of E onto itself. In particular, if U is a neighbourhood, so is αU for each $\alpha \neq 0$.*

Proof. If $f(x) = \alpha x = y$, then $f^{-1}(y) = x = \alpha^{-1}y$. Thus f is bicontinuous and so a homeomorphism.

PROPOSITION 3. *If $\mathcal{U}$ is a base of neighbourhoods (of the origin) then, for each $U \in \mathcal{U}$,*

(i) *U is absorbent;*
(ii) *there exists $V \in \mathcal{U}$ with $V+V \subseteq U$;*
(iii) *there is a balanced neighbourhood $W \subseteq U$.*

Proof. (i) If $a \in E$ and $f(\lambda) = \lambda a$, f is continuous at $\lambda = 0$ and so there is a neighbourhood $\{\lambda: |\lambda| \leqslant \epsilon\}$ of 0 mapped into U. Thus $\lambda a \in U$ for $|\lambda| \leqslant \epsilon$ and so $a \in \mu U$ for $|\mu| \geqslant \epsilon^{-1}$.

(ii) If $g(x, y) = x+y$, g is continuous at $x = o$, $y = o$ and so there are neighbourhoods V_1 and V_2 with $x+y \in U$ for $x \in V_1$ and $y \in V_2$. There exists $V \in \mathcal{U}$ with $V \subseteq V_1 \cap V_2$; then $V+V \subseteq U$.

(iii) If $h(\lambda, x) = \lambda x$, h is continuous at $\lambda = 0$, $x = o$ and so there exist a neighbourhood V and $\epsilon > 0$ with $\lambda x \in U$ for $|\lambda| \leqslant \epsilon$ and $x \in V$. Hence $\lambda V \subseteq U$ for $|\lambda| \leqslant \epsilon$ and so $\epsilon V \subseteq \mu U$ for $|\mu| \geqslant 1$. Thus $\epsilon V \subseteq W = \bigcap_{|\mu| \geqslant 1} \mu U$. But ϵV is a neighbourhood (Prop. 2) and so W is also a neighbourhood. If $x \in W$ and $0 < |\lambda| \leqslant 1$ then, if $|\mu| \geqslant 1$, $x \in (\mu/\lambda) U$ and so $\lambda x \in \mu U$ for $|\lambda| \leqslant 1$. Hence $\lambda x \in W$, so that W is balanced and clearly is contained in U.

It follows from Proposition 3 that every topological vector space has a base of balanced neighbourhoods. In the most important and useful topological vector spaces there is also a base of convex neighbourhoods of the origin. Such a space is called a *locally convex topological vector space*; we contract this to *convex space*.

THEOREM 2. *A convex space E has a base $\mathscr{U}$ of neighbourhoods of the origin with the following properties:*

C 1: *if $U \in \mathscr{U}$, $V \in \mathscr{U}$, there is a $W \in \mathscr{U}$ with $W \subseteq U \cap V$;*

C 2: *if $U \in \mathscr{U}$ and $\alpha \neq 0$, $\alpha U \in \mathscr{U}$;*

C 3: *each $U \in \mathscr{U}$ is absolutely convex and absorbent.*

Conversely, given a (non-empty) set $\mathscr{U}$ of subsets of a vector space E with the properties C 1–C 3, *there is a topology making E a convex space with $\mathscr{U}$ as a base of neighbourhoods.*

Proof. If E is a convex space, there is, by definition, a base of convex neighbourhoods. If U is one of them, $\bigcap_{|\mu| \geqslant 1} \mu U$ is a balanced neighbourhood contained in U (Prop. 3); it is also convex, being an intersection of convex sets. There is therefore a base $\mathscr{V}$ of absolutely convex neighbourhoods. Then the set $\mathscr{U}$ of all sets αV with $\alpha \neq 0$ and $V \in \mathscr{V}$ is the required base. For the sets of $\mathscr{U}$ are neighbourhoods, by Proposition 2, and $\mathscr{U}$ is a base; then N 2 implies C 1, C 2 follows from the construction of $\mathscr{U}$ and C 3 also from this and from Proposition 3.

Conversely, suppose that $\mathscr{U}$ is a set with the properties C 1–C 3. Let $\mathscr{V}$ be the set of all subsets of E that contain a set of $\mathscr{U}$, and, for each $a \in E$, take $\mathscr{V} + a$ to be the set of neighbourhoods of a. We have only to plod through the details of verifying that N 1–N 4 are satisfied and that the topology so defined is compatible with the algebraic structure of E. First, N 1–N 3 are straightforward;

for N 4, if $V \in \mathscr{V}$, there is a $U \in \mathscr{U}$ with $U \subseteq V$ and then it is easy to show that $V+a$ is a neighbourhood of every point of $\frac{1}{2}U+a$. To prove the continuity of addition at $x = a$, $y = b$, let $U \in \mathscr{U}$; then, if $x \in \frac{1}{2}U+a$ and $y \in \frac{1}{2}U+b$, $x+y \in U+a+b$. Finally, to prove that λx is continuous at $\lambda = \alpha$, $x = a$, it is enough to find η and δ so that $\lambda x - \alpha a \in U$ whenever $|\lambda - \alpha| < \eta$ and $x \in \delta U + a$. Now there is a $\mu > 0$ with $a \in \mu U$; choose η so that $0 < 2\eta < \mu^{-1}$ and then δ so that $0 < 2\delta < (|\alpha|+\eta)^{-1}$. Then

$$\lambda x - \alpha a = \lambda(x-a) + (\lambda - \alpha)\, a \in (|\alpha|+\eta)\,\delta U + \eta\mu U \subseteq U.$$

COROLLARY. *Let $\mathscr{V}$ be any set of absolutely convex absorbent subsets of a vector space E. Then there is a coarsest topology on E compatible with the algebraic structure in which every set in $\mathscr{V}$ is a neighbourhood. Under this topology E is a convex space and a base of neighbourhoods is formed by the sets*

$$\epsilon \bigcap_{1 \leqslant i \leqslant n} V_i \quad (\epsilon > 0, V_i \in \mathscr{V}).$$

Proof. The set $\mathscr{U}$ of subsets of the form $\epsilon \bigcap_{1 \leqslant i \leqslant n} V_i$ $(\epsilon > 0, V_i \in \mathscr{V})$ satisfies the conditions C1–C3 and so $\mathscr{U}$ is a base of neighbourhoods in a topology ξ on E making E a convex space. Also in any compatible topology in which the sets of $\mathscr{V}$ are neighbourhoods, the sets of $\mathscr{U}$ must also be neighbourhoods (by N 2 and Prop. 2); thus ξ is the coarsest such topology.

From now onwards we shall study only convex spaces. It is mainly these that occur in practice; this is fortunate because they have a richer theory, for a reason that we shall see in the next chapter. Occasionally there will be results that do not depend upon local convexity; we shall give proofs in the general case as long as this does not cause unreasonable complication.

PROPOSITION 4. *In a topological vector space, the closure of a convex set is convex, the closure of a balanced set is balanced and the closure of an absolutely convex set is absolutely convex.*

Proof. Let A be an absolutely convex set, let $a \in \bar{A}$, $b \in \bar{A}$ and $|\lambda| + |\mu| \leqslant 1$. For any neighbourhood U, there is a balanced neighbourhood V with $V + V \subseteq U$ (Prop. 3); then there are points $x \in A \cap (a+V)$ and $y \in A \cap (b+V)$ and so

$$\begin{aligned}\lambda x + \mu y &\in (\lambda A + \mu A) \cap (\lambda a + \mu b + \lambda V + \mu V)\\ &\subseteq A \cap (\lambda a + \mu b + V + V)\\ &\subseteq A \cap (\lambda a + \mu b + U).\end{aligned}$$

Hence $\lambda a + \mu b \in \bar{A}$, and thus $\bar{A}$ is absolutely convex. The proof is similar for a convex set and for a balanced set.

COROLLARY. *A topological vector space has a base of closed balanced neighbourhoods; further, a convex space has a base of closed neighbourhoods with the properties* C 1–C 3.

Proof. The closures of the sets in a base $\mathscr{U}$ of balanced neighbourhoods form a base of neighbourhoods. For if $U \in \mathscr{U}$, there is a neighbourhood $V \in \mathscr{U}$ with $V + V \subseteq U$; then if $x \in \bar{V}$, $x + V$ meets V and so $x \in V - V = V + V \subseteq U$. Thus any topological vector space has a base of closed balanced neighbourhoods. For a convex space, take $\mathscr{U}$ to be the base of Theorem 2; the proposition ensures that the closures of the sets of $\mathscr{U}$ remain absolutely convex and the other properties are immediate.

PROPOSITION 5. *If $\mathscr{U}$ is a base of neighbourhoods in the topological vector space E, then E is separated if and only if*

$$\bigcap_{U \in \mathscr{U}} U = \{o\}.$$

Proof. If E is separated and $x \neq o$, there is some $U \in \mathscr{U}$ with $x \notin U$ and so

$$\bigcap_{U \in \mathscr{U}} U = \{o\}.$$

Conversely, if this last condition holds and $x \neq y$ there is some U with $x - y \notin U$. By Proposition 3 there is a balanced neighbourhood V with $V + V \subseteq U$. Then $x + V$ and $y + V$ are disjoint neighbourhoods of x and y, for if $z \in (x+V) \cap (y+V)$ then

$$x - y = (z-y) - (z-x) \in V - V = V + V \subseteq U.$$

Therefore E is separated.

4. Seminorms. Let E be a vector space over Φ. A non-negative (finite) real-valued function p defined on E is called a *seminorm* if it satisfies

S 1: $p(x) \geqslant 0$;
S 2: $p(\lambda x) = |\lambda| p(x)$;
S 3: $p(x+y) \leqslant p(x)+p(y)$
for all $x, y \in E$ and all $\lambda \in \Phi$. By S 2, $p(o) = 0$, but it may happen that $p(x) = 0$ for some $x \neq o$. In fact $p^{-1}(0)$ is a vector subspace of E. A seminorm p for which $p(x) = 0$ implies $x = o$ is called a *norm*.

It follows from S 3 that

$$p(x) \leqslant p(y)+p(x-y), \quad p(y) \leqslant p(x)+p(y-x),$$

and so $|p(x)-p(y)| \leqslant p(x-y)$, an inequality which is often useful. So also is

Lemma 2. *Suppose that p and q are two seminorms on E and that $q(x) \leqslant 1$ whenever $p(x) < 1$. Then $q(x) \leqslant p(x)$ for all $x \in E$.*

Proof. If not, there are $x \in E$, $\alpha > 0$ with $0 \leqslant p(x) < \alpha < q(x)$ and then $p(x/\alpha) < 1$ but $q(x/\alpha) > 1$.

Seminorms are connected with absolutely convex absorbent sets:

Proposition 6. (i) *Let p be a seminorm on E. Then for each $\alpha > 0$, the sets $\{x: p(x) < \alpha\}$ and $\{x: p(x) \leqslant \alpha\}$ are absolutely convex and absorbent.*

(ii) *To each absolutely convex absorbent subset A of E corresponds a seminorm p, defined by*

$$p(x) = \inf\{\lambda: \lambda > 0, x \in \lambda A\},$$

and with the property that

$$\{x: p(x) < 1\} \subseteq A \subseteq \{x: p(x) \leqslant 1\}.$$

Proof. (i) This results from S 2, S 3 and the finiteness of p. (ii) The fact that A is absorbent ensures that $p(x)$ is finite for each $x \in E$. The conditions S 1 and S 2 are clearly satisfied; for S 3, let $x \in \lambda A$ and $y \in \mu A$ with $\lambda > 0$, $\mu > 0$. Then

$$x+y \in \lambda A+\mu A = (\lambda+\mu)A \quad \text{and so} \quad p(x+y) \leqslant \lambda+\mu.$$

Hence $p(x+y) \leqslant p(x)+p(y)$. The last part follows from the definition of p.

The seminorm p corresponding to the absolutely convex absorbent set A in this way is called the *gauge* of A. If also q is the gauge of the absolutely convex absorbent set B, then, straight from the definition of gauge:

(i) if $\alpha \neq 0$, the gauge of αA is $|\alpha|^{-1}p$,

(ii) the gauge of $A \cap B$ is $\sup\{p, q\}$,

(iii) if $A \subseteq B$ then $q(x) \leqslant p(x)$ for all $x \in E$.

If p is the gauge of A it is also the gauge of every absolutely convex absorbent set B satisfying

$$\{x: p(x) < 1\} \subseteq B \subseteq \{x: p(x) \leqslant 1\}.$$

The connection between seminorms and absolutely convex absorbent sets suggests that it should be possible to describe the topology of a convex space in terms of seminorms. The basic continuity properties required are collected together in:

PROPOSITION 7. (i) *In a convex space E, a seminorm p is continuous if (and only if) it is continuous at the origin.*

(ii) *If p is the gauge of the absolutely convex absorbent set U, p is continuous if and only if U is a neighbourhood. In this case the interior of U is $\{x: p(x) < 1\}$ and the closure of U is $\{x: p(x) \leqslant 1\}$.*

Proof. (i) If p is continuous at the origin and $\epsilon > 0$ is given, there is a neighbourhood V with $p(x) < \epsilon$ for $x \in V$. Now, if a is any point of E, $|p(x)-p(a)| \leqslant p(x-a) < \epsilon$ for $x \in a+V$.

(ii) If U is a neighbourhood and $\epsilon > 0$ is given, then $p(x) \leqslant \epsilon$ for $x \in \epsilon U$ and so p is continuous at the origin, and, by (i), on E also.

If p is continuous then $V = \{x: p(x) < 1\}$ is open, being the inverse image of the open interval $]-1, 1[$. But $V \subseteq U$ and so U is a neighbourhood. Now $\overline{V} = \{x: p(x) \leqslant 1\}$. For this latter set is closed and contains V; if x is a point of this set and W is any neighbourhood, then, since W is absorbent, there is some μ with $0 < \mu < 1$ and $-\mu x \in W$. Hence

$$(1-\mu)x \in x+W \quad \text{and} \quad p((1-\mu)x) = (1-\mu)p(x) \leqslant 1-\mu < 1.$$

Thus $(1-\mu)x \in V$ and so $x+W$ meets V. Hence $x \in \overline{V}$. Next, interior $\overline{V} = V$. For if $x \in$ interior $\overline{V}$, there is a neighbourhood W with $x+W \subseteq \overline{V}$. Then there is some μ with $0 < \mu < 1$ and $\mu x \in W$, and so $(1+\mu)x \in \overline{V}$. Hence $p((1+\mu)x) \leqslant 1$ and so $p(x) < 1$. Thus $x \in V$. Finally, $V \subseteq U \subseteq \overline{V}$ and so interior $U = V$ and $\overline{U} = \overline{V}$.

Proposition 7 shows that conditions exactly parallel to C 1, C 2 and C 3 in Theorem 2 and an exactly parallel theorem can be formulated in terms of seminorms. What is more useful, and indeed is the most usual way of setting up a convex space topology in applications, is the following:

THEOREM 3. *Given any set Q of seminorms on a vector space E, there is a coarsest topology on E compatible with the algebraic structure in which every seminorm in Q is continuous. Under this topology E is a convex space and a base of closed neighbourhoods is formed by the sets*

$$\{x: \sup_{1 \leqslant i \leqslant n} p_i(x) \leqslant \epsilon\} \quad (\epsilon > 0, p_i \in Q).$$

Proof. This follows from the Corollary of Theorem 2 and Proposition 7. For if p_i is the gauge of the absolutely convex neighbourhood V_i, then $\epsilon^{-1} \sup_{1 \leqslant i \leqslant n} p_i$ is the gauge of $\epsilon \bigcap_{1 \leqslant i \leqslant n} V_i$.

We say that the topology of E is *determined* by the set Q of seminorms.

PROPOSITION 8. *Under the topology determined by the set Q of seminorms, E is separated if and only if for each non-zero $x \in E$ there is some $p \in Q$ with $p(x) > 0$.*

Proof. This condition is easily seen to be equivalent to that in Proposition 5.

It follows that if E is a non-separated convex space, there are points $x \neq o$ with $p(x) = 0$ for all continuous seminorms p. The set of all such x is a vector subspace N of E; if $\mathscr{U}$ is a base of absolutely convex neighbourhoods, then $N = \bigcap_{U \in \mathscr{U}} U$, and N is the closure of the set consisting of the origin alone. (For $x \in N$ if and only if x belongs to every neighbourhood U of the origin and so if and only if the origin belongs to every neighbourhood $x + U$ of x.)

However, the convex spaces that occur in practice are nearly always separated, or if they are not, can be converted to separated spaces by the identification of elements whose difference lies in N (see Ch. v, Suppl. 2). For these reasons, we shall feel at liberty to restrict ourselves to separated spaces whenever this clarifies the theory or avoids the labouring of technical details.

If p is a norm on E, the topology on E determined by putting $Q = \{p\}$ is clearly separated. The space E is called *normable* if its topology can be defined from a norm p. The scalar field itself, under its natural topology, is normable; $p(\lambda) = |\lambda|$ is a suitable norm. In a general normed space, the norm of the point x is usually denoted thus: $\|x\|$. The set $U = \{x: \|x\| \leqslant 1\}$ is then called the *unit ball* and the sets ϵU ($\epsilon > 0$) form a base of neighbourhoods. A normed space is clearly metrisable; in fact

$$d(x, y) = \|x-y\|$$

is a suitable metric. More generally:

THEOREM 4. *The convex space E is metrisable if and only if it is separated and there is a countable base of neighbourhoods (of the origin). The topology of a metrisable space can always be defined by a metric that is invariant under translation.*

Proof. If E is metrisable it is certainly separated and has a countable base of neighbourhoods of the origin.

If E has a countable base, each neighbourhood contains an absolutely convex neighbourhood and so there is a base (U_n) of absolutely convex neighbourhoods. Let p_n be the gauge of U_n. Put

$$f(x) = \sum_{n=1}^{\infty} 2^{-n} \inf\{p_n(x), 1\}.$$

Then $f(x+y) \leqslant f(x)+f(y)$, $f(-x) = f(x)$ and if $f(x) = 0$, $p_n(x) = 0$ for all n and so $x = o$ since E is separated. Define d by

$$d(x, y) = f(x-y);$$

then d is a metric and $d(x+z, y+z) = d(x, y)$, so that d is invariant under translation. In the metric topology, the sets

$$V_n = \{x: f(x) < 2^{-n}\}$$

form a base of neighbourhoods. But V_n is open in the original topology, since each p_n, and so f, is continuous; also $V_n \subseteq U_n$, for if $x \notin U_n$ then $p_n(x) \geqslant 1$ and so $f(x) \geqslant 2^{-n}$. Hence d defines the original topology on E.

COROLLARY. *If the topology on the separated space E is the coarsest convex topology making a sequence of absolutely convex* *absorbent sets neighbourhoods (or a sequence of seminorms continuous), then E is metrisable.*

For, in the coarsest convex topology making the sets V_n neighbourhoods, the sets $s^{-1} \bigcap_{1 \leqslant i \leqslant r} V_{n_i}$ (r and s positive integers) form a neighbourhood base (Th. 2, Corollary), which is countable.

The function f constructed in the proof of Theorem 4 fails to be a norm only because $f(\lambda x)$ is not equal to $|\lambda| f(x)$. In fact there are convex metrisable spaces that are not normable (see, for example, Suppl. 2*b*), including many that are of practical importance.

SUPPLEMENT

In this section we give examples of convex spaces.

(1) *Finite-dimensional spaces.* Euclidean n-dimensional space, with points $x = (x_1, x_2, \ldots, x_n)$, becomes a convex space under the topology defined by the norm

$$\|x\| = \sqrt{\sum_{1 \leqslant i \leqslant n} |x_i|^2}.$$

This topology is in fact the only one compatible with the algebraic structure (see Ch. II, Prop. 11). In particular, the scalar field itself is a one-dimensional Euclidean space.

(2) *Spaces of continuous functions.* (*a*) The set of all real (or complex) valued functions continuous on a bounded closed interval $[a, b]$ becomes a convex space under the norm

$$\|x\| = \sup_{a \leqslant t \leqslant b} |x(t)|.$$

(Here and later $x(t)$ denotes the value of the function x at the point t.) This is a specific example of a class of normed spaces of continuous functions. Let $\mathscr{C}(S)$ be the set of real (or complex) valued functions continuous on the topological space S. If S is compact, $\mathscr{C}(S)$ is a normed space under the *topology of uniform convergence on* S, i.e. the topology defined by the norm

$$\|x\| = \sup_{t \in S} |x(t)|.$$

(*b*) The set of all real (or complex) valued functions continuous on $]-\infty, \infty[$ forms a convex space under the topology determined by the seminorms

$$p_n(x) = \sup_{-n \leqslant t \leqslant n} |x(t)| \quad (n = 1, 2, \ldots).$$

This space is clearly separated and metrisable (Th. 4) but not normable. For if there were a norm defining the same topology then the unit ball U would contain a neighbourhood

$$V = \{x: p_n(x) < \epsilon\}$$

for some n and $\epsilon > 0$; then $p_n(x) = 0$ implies $\|x\| = 0$ and so $x = o$. But for each n there is certainly a continuous function x that is not identically zero but vanishes on $[-n, n]$, so that $p_n(x) = 0$.

Again, this space is a specific example of a space $\mathscr{C}(S)$ of real (or complex) valued functions continuous on the topological space S, this time with the *topology of compact convergence on* S. For each compact subset A of S, let

$$p_A(x) = \sup_{t \in A} |x(t)|;$$

the set of seminorms p_A determines a convex space topology on $\mathscr{C}(S)$. If S is a separated locally compact space, the union of a sequence of compact sets, then $\mathscr{C}(S)$ is metrisable. (For S is the union of a sequence (A_n) of compact sets with each A_n contained in the interior of A_{n+1}; the seminorms corresponding to the sets A_n determine the topology, since any compact set is then contained in some A_n.)

(c) Let $\mathscr{K}(S)$ be the set of real (or complex) valued functions continuous and of compact support on the separated locally compact space S. (The *support* of a function is the smallest closed set outside which it vanishes.) This set can be given the topology of uniform convergence, with norm

$$\|x\| = \sup_{t \in S} |x(t)|,$$

or the (coarser) topology of compact convergence, or another topology, which is especially relevant to integration theory. For each compact subset A of S, let $\mathscr{K}_A(S)$ be the vector subspace of $\mathscr{K}(S)$ consisting of the functions with supports contained in A. Let $\mathscr{U}$ be the set of all absolutely convex absorbent subsets U of

$\mathscr{K}(S)$ such that, for every compact A, $U \cap \mathscr{K}_A(S)$ is a neighbourhood in $\mathscr{K}_A(S)$ under the topology of uniform convergence on A. Then $\mathscr{U}$ is a base of neighbourhoods for a convex topology on $\mathscr{K}(S)$, which is in fact the finest convex topology inducing on each $\mathscr{K}_A(S)$ a topology coarser than its uniform convergence topology. This is an inductive limit topology (see Ch. v, § 2). It is finer than the topology of compact convergence or the topology of uniform convergence; under each, $\mathscr{K}(S)$ is separated. When S is $]-\infty, \infty[$, the topology is not metrisable (Ch. vii, Prop. 5). (See also Ch. ii, Suppl. 2; Ch. iii, Suppl. 1.)

(3) *Spaces of indefinitely differentiable functions* (i.e. functions that can be differentiated any number of times). (*a*) The set of all real (or complex) valued indefinitely differentiable functions on the interval $[a, b]$ becomes a metrisable convex space under the topology defined by the seminorms

$$p_m(x) = \sup_{a \leqslant t \leqslant b} |x^{(m)}(t)| \quad (m = 0, 1, \ldots).$$

(*b*) Let $\mathscr{E}$ be the set of all real (or complex) valued indefinitely differentiable functions on $]-\infty, \infty[$. Under the *topology of compact convergence for all derivatives* defined by the seminorms

$$p_{mn}(x) = \sup_{-n \leqslant t \leqslant n} |x^{(m)}(t)| \quad (m = 0, 1, \ldots; n = 1, 2, \ldots),$$

$\mathscr{E}$ is a metrisable convex space.

(*c*) Let $\mathscr{S}$ be the set of all real (or complex) valued indefinitely differentiable functions on $]-\infty, \infty[$ with the property that, for all integers $m \geqslant 0$ and $n \geqslant 0$, $|t|^n x^{(m)}(t) \to 0$ as $|t| \to \infty$ (functions of *rapid decrease*). With the topology determined by the seminorms

$$p_{mn}(x) = \sup_{]-\infty, \infty[} |(1 + |t|^n) x^{(m)}(t)|,$$

$\mathscr{S}$ becomes a metrisable convex space; the topology is finer than that induced on it as a subspace of $\mathscr{E}$ under the topology of com pact convergence for all derivatives.

(*d*) Let $\mathscr{D}$ be the set of all real (or complex) valued indefinitely differentiable functions of compact support on $]-\infty, \infty[$. (There are such functions, e.g. if $a < b < c$, put

$$x(t) = \exp\left(1 - \frac{(c-b)(b-a)}{(c-t)(t-a)}\right)$$

on $]a, c[$ and zero elsewhere; then $x(b) = 1$ and $x(t)$ vanishes, together with all its derivatives, outside $]a, c[$.) To topologise $\mathcal{D}$, let $\mathcal{D}_n$ be the vector subspace of $\mathcal{D}$ consisting of the functions with support contained in $[-n, n]$ and let $\mathcal{D}_n$ have the metrisable topology of (a). Take as a base of neighbourhoods in $\mathcal{D}$ the set of all absolutely convex absorbent subsets that intersect every $\mathcal{D}_n$ in a neighbourhood. This is a non-metrisable inductive limit topology similar to that in 2*c*. It is finer than those induced on $\mathcal{D}$ as a subspace of $\mathcal{E}$ or even of $\mathcal{S}$ (for the topologies of (b) and (c) induce on each $\mathcal{D}_n$ a topology coarser than that in (a)). Under the topology of compact convergence for all derivatives $\mathcal{D}$ (and so also $\mathcal{S}$) is dense in $\mathcal{E}$.

The spaces $\mathcal{D}$, $\mathcal{S}$ and $\mathcal{E}$ arise in the theory of distributions; more generally, instead of being defined on the real line $R =]-\infty, \infty[$, the functions may be defined on any finite-dimensional space R^n. (See also Ch. II, Suppl. 3; Ch. III, Suppl. 1; Ch. IV, Suppl. 2; Ch. VII, Suppl. 1.)

(4) *Spaces of holomorphic functions.* Let D be a domain in the complex plane and $\mathcal{H}(D)$ the set of all functions holomorphic on D, with the topology of compact convergence on D (cf. 2*b*). If

$$A_n = \{t\colon |t| \leqslant n,\, d(t, \sim D) \geqslant n^{-1}\},$$

the seminorms $$p_n(x) = \sup_{t \in A_n} |x(t)|$$

determine the topology. For each A_n is compact; if A is a compact subset of D, $d(A, \sim D) > 0$ and so there is some n with $A \subseteq A_n$. Hence $\mathcal{H}(D)$ is metrisable. On $\mathcal{H}(D)$ this topology is in fact the same as that of compact convergence for all derivatives (cf. 3*b*). It is clearly coarser; on the other hand, if $t \in A_n$ and Γ is the circle $|\tau - t| = (2n)^{-1}$,

$$|x^{(m)}(t)| = \left|\frac{m!}{2\pi i}\int_\Gamma \frac{x(\tau)\, d\tau}{(\tau - t)^{m+1}}\right| \leqslant m!(2n)^m p_{2n}(x)$$

and so it is also finer. (See also Ch. II, Suppl. 4; Ch. III, Suppl. 1; Ch. IV, Suppl. 2.)

(5) *Sequence spaces.* The vector space of all sequences $x = (x_n)$ of scalars, and various vector subspaces of it, can be topologised (see Ch. II, Suppl. 5). Examples of normed sequence spaces are:

c, the space of all convergent sequences with $\|x\| = \sup |x_n|$;

c_0, the space of all sequences convergent to zero, with the same norm;

l^p $(p \geqslant 1)$, the space of all sequences for which

$$\|x\| = \left\{ \sum_{n=1}^{\infty} |x_n|^p \right\}^{1/p} < \infty;$$

m $(= l^\infty)$, the space of all bounded sequences with $\|x\| = \sup |x_n|$.

If $0 < p < 1$, the space l^p of all sequences for which

$$\sum_{n=1}^{\infty} |x_n|^p < \infty$$

can be topologised with the metric

$$d(x, y) = \sum_{n=1}^{\infty} |x_n - y_n|^p,$$

and this topology is compatible with the algebraic structure. However, it is not locally convex. (For if it were, $\{x: d(x, o) \leqslant 1\}$ would contain an absolutely convex neighbourhood U, which in turn would contain a set of the form $\{x: d(x, o) \leqslant \epsilon\}$ for some $\epsilon > 0$. Then if $x^{(r)} = (x_n^{(r)})$ with $x_n^{(r)} = 1$ for $n = r$ and zero otherwise, $\epsilon^{1/p} x^{(r)} \in U$ and so

$$y = \epsilon^{1/p} s^{-1} \sum_{1 \leqslant r \leqslant s} x^{(r)} \in U.$$

But $d(y, o) = \epsilon s^{1-p} > 1$ for sufficiently large s.)

(6) *Integration spaces.* If S is a measure space with measure μ, the vector space of all measurable functions $x = x(t)$ on S, and various subspaces of it, can be topologised (see Ch. II, Suppl. 6). Examples of normed integration spaces are $\mathscr{L}^p$ $(p \geqslant 1)$, the space of 'functions' with

$$\|x\| = \left\{ \int_S |x(t)|^p \, d\mu \right\}^{1/p} < \infty,$$

and $\mathscr{M}$ $(= \mathscr{L}^\infty)$, the space of 'functions' with

$$\|x\| = \text{ess.}\sup |x(t)| < \infty.$$

To make these into genuine normed spaces, we have to take the elements to be equivalence classes of functions, writing $x = y$ for $x(t) = y(t)$ almost everywhere on S, an example of the process of identification referred to in §4 for ensuring that the space is separated. If $0 < p < 1$, the space $\mathscr{L}^p$ (of all 'functions' with

$$\int_S |x(t)|^p \, d\mu < \infty)$$

with the metric
$$d(x,y) = \int_S |x(t) - y(t)|^p \, d\mu$$

is another example of a topological vector space that is not locally convex. (See also Ch. II, Suppl. 6; Ch. III, Suppl. 1.)

(7) *Topology of pointwise convergence.* If S is any set, the vector space of all real (or complex) valued functions on S can be made into a convex space by giving it the topology of pointwise convergence, determined by the seminorms $p_t(x) = |x(t)|$ for each $t \in S$. (See also Ch. V, Suppl. 1.)

(8) *Finest convex topology.* Any vector space E can be made into a convex space by taking as a base of neighbourhoods of the origin the set of all absolutely convex absorbent subsets. This is the finest convex topology on E; every seminorm on E is continuous, and E is separated (for if $x \neq o$, extend $\{\frac{1}{2}x\}$ to form a base A of E; the absolutely convex envelope of A is absorbent and does not contain x). (See also Ch. II, Suppl. 7; Ch. III, Suppl. 2; Ch. IV, Suppl. 3; Ch. V, Suppl. 1; Ch. VI, Suppl. 1.)

CHAPTER II

DUALITY AND THE HAHN–BANACH THEOREM

In this chapter we begin the study of duality, which plays a central part in the development of the theory of convex spaces. The dual of a convex space is the vector space of all continuous linear mappings of the space into the scalar field. The main result of the chapter is the Hahn–Banach theorem, which is concerned with the existence of such linear mappings; we give a number of different forms of this important theorem. In particular, the Hahn–Banach theorem shows that a separated convex space and its dual are in a symmetric relation to one another, both algebraically and topologically, by means of their weak topologies. Some further consequences of the Hahn–Banach theorem given in the later sections, and in particular Theorem 4, will be of constant use in the chapters that follow.

1. Linear mappings. Let E and F be two vector spaces over the same field Φ (of real or complex numbers). The mapping f of E into F is called *linear* if

$$f(x+y) = f(x)+f(y), \quad f(\lambda x) = \lambda f(x),$$

for all $x \in E$, $y \in E$, and $\lambda \in \Phi$. A linear mapping is sometimes called a linear *operator*, or *operation*, or *transformation*. The linear mapping f is (1, 1) if and only if $f^{-1}(o) = \{o\}$; in general $f^{-1}(o)$ is a vector subspace of E.

In the set L of all linear mappings of E into F, addition and multiplication by scalars can be defined by

$$(f+g)(x) = f(x)+g(x), \quad (\lambda f)(x) = \lambda(f(x));$$

then L becomes a vector space over Φ.

When E and F are both topological vector spaces, the continuous linear mappings of E into F form a vector subspace of L, because the continuity of f and g imply the continuity of $f+g$ and λf. There is a simple criterion for continuity of linear mappings:

PROPOSITION 1. *If E and F are topological vector spaces and f is a linear mapping of E into F, then f is continuous on E if (and only if) f is continuous at the origin.*

Proof. If f is continuous at o, and V is any neighbourhood in F, there is a neighbourhood U in E with $f(U) \subseteq V$. Then for each point a of E, $f(a+U) = f(a)+f(U) \subseteq f(a)+V$, and so f is continuous at a.

COROLLARY. *If E and F are normed spaces and f is a linear mapping of E into F, then f is continuous if and only if there is a constant α with $\|f(x)\| \leqslant \alpha\|x\|$ for all $x \in E$.*

For f is continuous at the origin if and only if there is some α with $\|f(x)\| \leqslant \alpha$ whenever $\|x\| \leqslant 1$. Hence (Ch. I, Lemma 2) $\|f(x)\| \leqslant \alpha\|x\|$ for all $x \in E$.

Let T be any subset of L. Saying that each $f \in T$ is continuous is equivalent to:

for each $f \in T$ and each neighbourhood V in F, there is a neighbourhood U_f with $f(U_f) \subseteq V$.

If the same neighbourhood U in E will suffice for all the f in T, the set T is called *equicontinuous*. Thus T is equicontinuous if:

for each neighbourhood V in F, there is a neighbourhood U in E with $f(U) \subseteq V$ for all $f \in T$,

or equivalently, for each neighbourhood V in F, $\bigcap_{f \in T} f^{-1}(V)$ is a neighbourhood in E. When E and F are normed spaces, then T is equicontinuous if and only if there is a constant α with $\|f(x)\| \leqslant \alpha\|x\|$ for all $f \in T$.

Let E and F be two topological vector spaces and let f be a linear mapping of E into F. If f is (1, 1) and maps E onto F, then f^{-1} is a linear mapping of F onto E. If also f is bicontinuous, f is called an *isomorphism* of E onto F and the topological vector spaces E and F are called *isomorphic*. When E and F are normed spaces, it follows from the Corollary to Proposition 1 that a linear mapping f of E onto F is an isomorphism if and only if there are constants $\alpha > 0$ and β with $\alpha\|x\| \leqslant \|f(x)\| \leqslant \beta\|x\|$.

2. Linear forms and the Hahn–Banach theorem. If E is a vector space over Φ, a linear mapping of E into the scalar field Φ

itself is called a *linear form* (or *linear functional*) on E. The set of all linear forms on E is a vector space over Φ called the *algebraic dual* (or *algebraic conjugate*) of E and denoted by E^*. There are always enough linear forms in E^* to distinguish the elements of E, in the following sense.

PROPOSITION 2. *For each non-zero* $a \in E$ *there is a linear form* $f \in E^*$ *with* $f(a) \neq 0$.

Proof. By Theorem 1 of Chapter I, $\{a\}$ can be extended to form a base A of E; define f by putting, for example, $f(x) = 1$ on A and extending to E by linearity.

If f is a non-zero linear form on E, then $H = f^{-1}(0)$ is a proper vector subspace of E and the set $\{H, a\}$ spans E for every $a \notin H$ (if $x \in E$, then $x - (f(x)/f(a))\, a \in H$). Thus there is no proper vector subspace of E strictly containing H, or, in other words, H is a maximal proper vector subspace of E. Conversely, if H is a maximal proper vector subspace of E, there are linear forms f such that $f^{-1}(0) = H$. (For there is some $a \notin H$, and then any element of E is of the form $x + \lambda a$ with $x \in H$; define $f(x + \lambda a) = \lambda\alpha$, where $\alpha \neq 0$.) The maximal proper vector subspaces of E are called *hyperplanes* (through the origin).

When E is a topological vector space, the vector subspace of E^* consisting of those linear forms that are continuous is called the (continuous) *dual* (or *conjugate*) of E, and denoted by E'. In a general topological vector space it is possible for the only continuous linear form to be the zero form $f(x) = 0$ for all $x \in E$ (e.g. Suppl. 6). The crucial property of convex spaces is that nothing like this can happen to them.

A linear form f on E is continuous if and only if it is bounded on some neighbourhood. (For if $|f(x)| \leqslant \alpha$ on U, then $|f(x)| \leqslant \epsilon$ on $\epsilon\alpha^{-1}U$ and so f is continuous, by Proposition 1.) If E is a convex space and p is a continuous seminorm such that $|f(x)| \leqslant p(x)$ for all $x \in E$, then f is continuous, for it is bounded on the neighbourhood $\{x: p(x) \leqslant 1\}$. If f is continuous on E then $|f|$ is a continuous seminorm.

A non-zero linear form f is completely specified if its null-space $f^{-1}(0) = H$ is given and the value of f at one other point $a \notin H$ is known. For example, H and a point a with $f(a) = 1$ fix f.

LEMMA 1. *Let f be a linear form on a vector space, $H = f^{-1}(0)$, $f(a) = 1$ and $V = \{x: |f(x)| < 1\}$. Then if U is a balanced set, $(a+U) \cap H = \emptyset$ if and only if $U \subseteq V$.*

Proof. Suppose that $U \subseteq V$. Then if $x \in U$,

$$f(a+x) = 1+f(x) \neq 0$$

since $x \in V$, and so $a+U$ does not meet H. Conversely, suppose that $x \in U$ but that $|f(x)| \geqslant 1$. Then $y = -x/f(x) \in U$ and $f(a+y) = 0$, so that $(a+U) \cap H \neq \emptyset$.

THEOREM 1. *If f is a linear form on a topological vector space, then f is continuous if and only if $f^{-1}(0)$ is closed.*

Proof. If f is continuous then $f^{-1}(0)$, the inverse image by f of the closed subset $\{0\}$ of Φ, is closed.

Suppose that $H = f^{-1}(0)$ is closed and let $V = \{x: |f(x)| < 1\}$. If f is not the zero linear form (in which case it is continuous) there is some point a with $f(a) = 1$. Hence there is a balanced neighbourhood U with $(a+U) \cap H = \emptyset$. By Lemma 1, $U \subseteq V$, so that V is a neighbourhood and f, being bounded on V, is continuous.

The proof can be adapted to show that $f^{-1}(0)$ is either closed or dense in the whole space E, so that $\overline{f^{-1}(0)} = E$, but a more direct proof of this is based on:

PROPOSITION 3. *If M is a vector subspace of a topological vector space then so is $\overline{M}$.*

Proof. Let $x \in \overline{M}$, $y \in \overline{M}$ and let U be a neighbourhood. There is a neighbourhood V with $V+V \subseteq U$. Then $x+V$ and $y+V$ meet M and so $x+y+U$ meets $M+M = M$. Hence $x+y \in \overline{M}$. Similarly if $x \in \overline{M}$ then $\lambda x \in \overline{M}$ for all $\lambda \in \Phi$.

PROPOSITION 4. *In a topological vector space a hyperplane is either closed or dense.*

Proof. Let H be a hyperplane in the topological vector space E. By Proposition 3, $\overline{H}$ is a vector subspace. But $\overline{H}$ contains H and H is maximal; hence either $\overline{H} = H$ or $\overline{H} = E$.

The Hahn–Banach theorem, which asserts that a continuous linear form on a vector subspace of E has a continuous extension to the whole of E, is one of the fundamental theorems of functional analysis. There are many different forms, of slightly varying degrees of generality, from which the main extension result can be deduced. Here we choose a theorem of geometrical character. It is remarkable that the theorem is not trivial in a finite-dimensional space, or, what is equivalent, that one part of the difficulty lies in the proof that the linear form can be extended to a subspace only one dimension larger; Lemma 2 copes with the geometrical parallel to this. Another difficulty is that the cases of a vector space over the real field and over the complex field need separate treatment; Lemma 3 is designed to overcome this. Finally, in an infinite-dimensional space, rather than extending the linear form by one dimension at a time, we should expect a use of the maximal axiom; this is to be found in the proof of the theorem itself.

THEOREM 2. *Suppose that A is an open convex subset of a convex space and that M is a vector subspace not meeting A. Then there is a closed hyperplane containing M and not meeting A.*

First we prove:

LEMMA 2. *Suppose that E is a real convex space, A an open convex subset of E and H a vector subspace not meeting A. Then either H is a hyperplane or there is a point $x \notin H$ such that the vector subspace spanned by H and x does not meet A.*

Proof. Let $C = H + \bigcup_{\lambda>0} \lambda A$. Then C is open,

$$-C = H + \bigcup_{\lambda<0} \lambda A \quad \text{and} \quad C \cap (-C) = \emptyset.$$

For if $x \in C \cap (-C)$, then $x = h + \lambda a = h' - \lambda' a'$ for some $h, h' \in H$, $a, a' \in A$ and $\lambda, \lambda' > 0$, and so $\lambda a + \lambda' a' \in H$. But since A is convex, $\lambda a + \lambda' a' \in (\lambda + \lambda') A$ which does not meet H.

(i) Suppose that $H \cup C \cup (-C) \neq E$. Then there is some $x \notin H$ with $x \notin C \cup (-C)$. If the vector subspace spanned by H and x meets A, in y say, then for some $\lambda \neq 0$, $x \in \lambda y + H \subseteq C \cup (-C)$. Hence the vector subspace does not meet A.

(ii) Suppose that $H \cup C \cup (-C) = E$. If H is not a hyperplane there is some point $a \in C$ so that H and a together do not span E; hence there is some point $b \in -C$ not in the span of H and a. Let $f(\lambda) = (1-\lambda) a + \lambda b$ $(0 \leqslant \lambda \leqslant 1)$. Now f is continuous and

C is open and so $I = f^{-1}(C)$ and $J = f^{-1}(-C)$ are open in $[0, 1]$. Also $0 \in I$, $1 \in J$ and $I \cap J = \emptyset$ since $C \cap (-C) = \emptyset$. Let

$$\alpha = \sup\{\lambda : \lambda \in I\}.$$

Then

$$\alpha \in \bar{I} \cap \overline{(\sim I)} \subseteq \overline{(\sim J)} \cap \overline{(\sim I)} = (\sim J) \cap (\sim I).$$

Hence $f(\alpha) \notin C \cup (-C)$. Thus $f(\alpha) \in H$, that is, $(1-\alpha)a + \alpha b \in H$. But b is not in the span of H and a; therefore H is a hyperplane.

Lemma 3. *Suppose that E is a complex vector space and that H is a real hyperplane in E. Then $H \cap (iH)$ is a (complex) hyperplane.*

Proof. Suppose that $a \notin H \cap iH$, and suppose, for example, that $a \notin H$. Then $ia \notin iH$, which is a real hyperplane, and so $a = \alpha i a + b$ with α real and $b \in iH$; then $(1+\alpha i)b = (1+\alpha^2)\, a \notin H$, and so $b \notin H$. Now if $x \in E$, $x = \beta b + y$ with β real and $y \in H$, and then $y = \gamma i b + z$ with γ real and $z \in iH$. Hence $z \in H$; thus $x = (\beta + \gamma i)\, b + z = (\lambda + \mu i)\, a + z$, say, with $z \in H \cap iH$. Therefore $H \cap iH$ is a complex hyperplane.

Proof of Theorem 2. First consider the case when E is a real vector space. Let $\mathscr{E}$ be the set of vector subspaces of E containing M and not meeting A. Apply the maximal axiom to the (trivial) chain $\mathscr{C} = \{M\}$; there is a maximal chain $\mathscr{M}$ in E with $\mathscr{C} \subseteq \mathscr{M} \subseteq \mathscr{E}$. Let H be the union of the sets of $\mathscr{M}$. Then clearly H is a vector subspace of E not meeting A. By Lemma 2, it is a hyperplane, because the other possibility would contradict the maximality of $\mathscr{M}$ (we could add the span of $H \cup \{x\}$ to $\mathscr{M}$). Also H is closed, because otherwise it is dense in E (Prop. 4) and meets every open set, including A.

If E is a complex vector space, it is also a real vector space (simply by restricting the scalars to be real) and so there is a real closed hyperplane K containing M and not meeting A. Then $H = K \cap (iK)$ is a complex closed hyperplane containing $M \cap (iM) = M$ and not meeting A.

Corollary. *Every closed vector subspace of a convex space is the intersection of the closed hyperplanes containing it.*

Proof. Let a be a point not in the closed vector subspace M. Then there is an open convex neighbourhood A of a not meeting M.

By the theorem there is a closed hyperplane H containing M and not meeting A; thus $a \notin H$.

THEOREM 3. (Hahn–Banach extension theorem.) *Suppose that p is a seminorm on the vector space E and that f is a linear form on a vector subspace M of E with $|f(x)| \leqslant p(x)$ for all $x \in M$. Then there is a linear form f_1 on E extending f and with $|f_1(x)| \leqslant p(x)$ for all $x \in E$.*

Proof. Let E have the topology determined by the seminorm p and let $U = \{x : p(x) < 1\}$. Assuming f non-zero, take a in M with $f(a) = 1$ and let $A = a + U$. Let $N = f^{-1}(0)$; if $x \in U \cap M$ then $|f(x)| < 1$. By Lemma 1, applied in M, N does not meet $a + (U \cap M) = A \cap M$ and so does not meet the open convex set A. By Theorem 2 there is a closed hyperplane H containing N and not meeting A. Let f_1 be the linear form with $H = f_1^{-1}(0)$ and $f_1(a) = 1$. Then f_1 extends f; by Lemma 1 again, $p(x) < 1$ implies $|f_1(x)| < 1$ and so (Ch. I, Lemma 2) $|f_1(x)| \leqslant p(x)$.

COROLLARY 1. *Any continuous linear form defined on a vector subspace of a convex space E has a continuous extension to E.*

Proof. There is an absolutely convex neighbourhood U with $|f(x)| \leqslant 1$ on $U \cap M$; if p is the gauge of U, then $|f(x)| \leqslant p(x)$ on M. Hence f has an extension f_1 to E satisfying $|f_1(x)| \leqslant p(x)$ and so continuous.

COROLLARY 2. *If a is any point of the vector space E and p is a seminorm on E, there is a linear form f on E with $|f(x)| \leqslant p(x)$ and $f(a) = p(a)$.*

For on M, the vector subspace spanned by a, we can define $f(\lambda a) = \lambda p(a)$ and extend f to E.

COROLLARY 3. *Let E be a separated convex space with dual E'. If $f(a) = 0$ for all $f \in E'$, then $a = o$.*

Proof. If $a \neq o$, there is a continuous seminorm p with $p(a) > 0$ (Ch. I, Prop. 8). Hence, by Corollary 2, there is a continuous linear form f with $f(a) \neq 0$.

We now deduce some other useful forms of the Hahn–Banach theorem which are of combined analytic and geometric type.

PROPOSITION 5. (Hahn–Banach separation theorem.) *Let E be a convex space. Suppose that A and B are disjoint convex sets and* *that A is open. Then there is a continuous linear form f such that $f(A)$ and $f(B)$ are disjoint (f separates A and B).*

Proof. The set $A–B$ is open and convex and does not contain the origin. Hence, by Theorem 2, there is a closed hyperplane $H = f^{-1}(0)$, say, containing the vector subspace $\{o\}$ and not meeting $A–B$. The linear form f is continuous since H is closed, and $f(A)$ and $f(B)$ do not meet.

In drawing corollaries from this proposition we make use of a further algebraic property of linear forms:

Lemma 4. *Any non-zero linear form on E maps open subsets of E onto open sets of scalars.*

Proof. Let A be an open set and $x \in A$; then $A-x$ contains a neighbourhood and so is absorbent. If f is a non-zero linear form there is some $a \in E$ with $f(a) = 1$, and then there is some $\alpha > 0$ with $\lambda a \in A-x$ for $|\lambda| \leqslant \alpha$. Then $f(x)+\lambda \in f(A)$ for $|\lambda| \leqslant \alpha$. Hence $f(A)$ is open.

Corollary 1 of Proposition 5. *If B is a convex subset of a convex space and $a \notin \bar{B}$, then there is a continuous linear form f with $f(a) \notin \overline{f(B)}$.*

Proof. Since $a \notin \bar{B}$ there is an open absolutely convex neighbourhood U with $(a+U) \cap B = \emptyset$. By the proposition there is a continuous linear form f with $f(a+U) \cap f(B) = \emptyset$. But $f(a+U)$ is open (Lemma 4) and so $f(a) \notin \overline{f(B)}$.

Corollary 2. *If B is an absolutely convex subset of a convex space and $a \notin \bar{B}$, then there is a continuous linear form f with $|f(x)| \leqslant 1$ for all $x \in B$ and $f(a) > 1$.*

Proof. By Corollary 1 there is a continuous linear form g with $g(a) \notin \overline{g(B)}$; now $\overline{g(B)}$ is absolutely convex and so

$$\sup \{|g(x)|: x \in B\} < |g(a)|.$$

Let $\alpha = \sup\{|g(x)|: x \in B\}$ and put $f = (|g(a)|/\alpha g(a))\,g$ if $\alpha \neq 0$; if $\alpha = 0$ then $f = (2/g(a))\,g$ will suffice.

Corollary 3. *Let E be a real convex space. If A and B are disjoint convex subsets of E and A is open, then there is a continuous linear form f and a constant α with $f(x) > \alpha$ for all $x \in A$ and $f(x) \leqslant \alpha$ for all $x \in B$.*

Proof. By the proposition, there is a continuous linear form f such that $f(A)$ and $f(B)$ do not meet; since they are convex sets of real numbers we may suppose that

$$\sup\{f(x): x \in B\} \leqslant \inf\{f(x): x \in A\}$$

(by multiplying f by -1 if necessary). Let $\alpha = \inf\{f(x): x \in A\}$. Then $f(x) \leqslant \alpha$ for all $x \in B$. Also since $f(A)$ is open (Lemma 4), $\alpha \notin f(A)$ and so $f(x) > \alpha$ for $x \in A$.

PROPOSITION 6. *Let E be a real convex space. Suppose that f is a linear form on a vector subspace M of E and that $f(x) > 0$ on the (non-empty) intersection with M of an open convex set A. Then there is a linear form f_1, extending f, with $f_1(x) > 0$ on A.*

Proof. Let $N = f^{-1}(0)$. Then $N \cap A = \emptyset$ and so by Theorem 2 there is a hyperplane H containing N with $H \cap A = \emptyset$. Let $a \in A \cap M$ and define f_1 by $H = f_1^{-1}(0)$ and $f_1(a) = f(a)$. Then f_1 extends f and $f_1(x) > 0$ on A. For $f_1(a) = \lambda > 0$; if $x \in A$ and $f_1(x) = -\mu \leqslant 0$, then $y = (\lambda x + \mu a)/(\lambda + \mu) \in A$ since A is convex. Also $f_1(y) = 0$ and so $y \in H$. But $H \cap A = \emptyset$. Hence $f_1(x) > 0$ on A.

3. Duality and the weak topology. Let E be a convex space with (continuous) dual E'. Then E' is a vector subspace of the algebraic dual E^* of E. Also to each element x of E corresponds a linear form $\tilde{x}$ on E' defined by $\tilde{x}(f) = f(x)$. The mapping of E into E'^* thus defined is clearly linear; if E is separated it is also (1, 1), for $\tilde{x} = \tilde{y}$ if and only if $f(x) = f(y)$ for all $f \in E'$, and by Corollary 3 of Theorem 3 this is equivalent to $x = y$. Thus E is identified with a vector subspace $\tilde{E}$ of E'^*. We shall see that this algebraic symmetry between E and E', in which each is (isomorphic to) a vector subspace of the algebraic dual of the other, extends to a topological one; there are topologies on E' under which it is a separated convex space with (continuous) dual E.

It is convenient to use a notation that exhibits this symmetry; we denote the elements of E' by $x', y', \ldots$, and write $\langle x, x' \rangle$ for the value of the linear form x' at the point x of E. Then $\langle x, x' \rangle$ is a *bilinear form* on E and E' (for each fixed $x' \in E'$ it is a linear form on E and for each fixed $x \in E$ it is a linear form on E'), and the following two conditions are satisfied:

D: *for each* $x \neq o$ *in* E, *there is some* $x' \in E'$ *with* $\langle x, x' \rangle \neq 0$;

D′: *for each* $x' \neq o$ *in* E', *there is some* $x \in E$ *with* $\langle x, x' \rangle \neq 0$.

(D is Corollary 3 of Theorem 3; D′ is obvious.)

More generally, let E and E' be any two vector spaces over the same (real or complex) scalar field, and let $\langle x, x' \rangle$ be a bilinear form on E and E' satisfying the conditions D and D′. Then there is a natural linear mapping of E' into E^*, in which the image of $x' \in E'$ is the linear form f on E with $f(x) = \langle x, x' \rangle$. This mapping is (1, 1) because of D′, and so E' is (isomorphic to) a vector subspace of E^*. Similarly, D ensures that E is (isomorphic to) a vector subspace of E'^*. We then call (E, E') a *dual pair*.

Clearly if (E, E') is a dual pair, so is (E', E). We have shown that if E is a separated convex space with dual E', then (E, E') and so also (E', E) are dual pairs. For any vector space E with algebraic dual E^*, (E, E^*) is a dual pair (for D is a consequence of Proposition 2).

Let (E, E') be a dual pair. To each $x' \in E'$ corresponds a seminorm p on E defined by $p(x) = |\langle x, x' \rangle|$. The coarsest topology on E making all these seminorms continuous (cf. Ch. I, Th. 3) is called the *weak topology* on E determined by E', and denoted by $\sigma(E, E')$. It is clearly the coarsest topology on E under which all the linear forms in E' are continuous. In $\sigma(E, E')$, the sets

$$\{x \colon \sup_{1 \leqslant i \leqslant n} |\langle x, x'_i \rangle| \leqslant 1\} \quad (x'_i \in E')$$

form a base of (closed) neighbourhoods. (The ϵ of Chapter I, Theorem 3 can be replaced by 1 because $|\langle x, y' \rangle| \leqslant \epsilon$ if and only if $|\langle x, \epsilon^{-1} y' \rangle| \leqslant 1$.) The topology $\sigma(E, E')$ is convex and separated, because D ensures that the condition of Proposition 8 of Chapter I is satisfied.

The dual of E under $\sigma(E, E')$ certainly contains E'; we show that it is exactly equal to E', with the aid of an algebraic lemma.

LEMMA 5. *If* $f_0, f_1, \ldots, f_n$ *are linear forms on the vector space* E, *then either* f_0 *is a linear combination of* $f_1, f_2, \ldots, f_n$ *or there is an element* a *of* E *with* $f_0(a) = 1$ *and* $f_1(a) = f_2(a) = \ldots = f_n(a) = 0$.

Proof. Clearly we may suppose $f_1, f_2, \ldots, f_n$ linearly independent. If $n = 0$ there is nothing to prove. Suppose that the result is true for the integer $n-1$. Then for each i with $1 \leqslant i \leqslant n$, f_i is not a linear combination of $f_1, f_2, \ldots, f_{i-1}, f_{i+1}, \ldots, f_n$ and so by the induction hypothesis there are elements a_i with $f_i(a_j) = 0$ for $i \neq j$ and $f_i(a_i) = 1$. For each $x \in E$,

$$x - \sum_{1 \leqslant i \leqslant n} f_i(x)\, a_i \in \bigcap_{1 \leqslant i \leqslant n} f_i^{-1}(0) = N,$$

say. Then either there is an element $a \in N$ with $f_0(a) = 1$ (and $f_i(a) = 0$ for $1 \leqslant i \leqslant n$) or $f_0(y) = 0$ for all $y \in N$. In this case, for all $x \in E$,

$$f_0(x) = \sum_{1 \leqslant i \leqslant n} f_i(x) f_0(a_i) = \sum_{1 \leqslant i \leqslant n} \lambda_i f_i(x)$$

and so

$$f_0 = \sum_{1 \leqslant i \leqslant n} \lambda_i f_i.$$

Corollary. *If $f_1, f_2, \ldots, f_n$ are linearly independent linear forms on the vector space E, then there are elements $a_1, a_2, \ldots, a_n$ in E with $f_i(a_i) = 1$ and $f_i(a_j) = 0$ for $i \neq j$ $(1 \leqslant i \leqslant n,\ 1 \leqslant j \leqslant n)$.*

Proposition 7. *If (E, E') is a dual pair, then the dual of E under $\sigma(E, E')$ is E'.*

Proof. Let f be a linear form on E continuous under $\sigma(E, E')$. Then $|f(x)| \leqslant \alpha < 1$ on some neighbourhood of the form

$$U = \{x \colon \sup_{1 \leqslant i \leqslant n} |\langle x, x_i' \rangle| \leqslant 1\} \quad (x_i' \in E').$$

By Lemma 5, either f is a linear combination of $x_1', x_2', \ldots, x_n'$ or there is some $a \in E$ with $f(a) = 1$ but $\langle a, x_i' \rangle = 0$ for $1 \leqslant i \leqslant n$. But the latter would imply $a \in U$ and $|f(a)| > \alpha$; hence

$$f = \sum_{1 \leqslant i \leqslant n} \lambda_i x_i' \in E'.$$

Also each $x' \in E'$ is continuous under $\sigma(E, E')$ and so the dual of E under $\sigma(E, E')$ is E'.

If E is a separated convex space with dual E', then (E', E) is a dual pair and $\sigma(E', E)$ is a separated convex topology on E' under which the dual is E (the topology promised at the beginning of this section).

If (E, E') is a dual pair, any convex topology on E under which the dual is E' is called a *topology of the dual pair* (E, E'). Proposition 7 shows that $\sigma(E, E')$ is such a topology; it is the coarsest topology of the dual pair (E, E'). Since it is separated, they all are. Certain apparently topological properties depend only on the dual pair and not on the particular topology of the dual pair. This means that the study of such properties in a separated convex space can be carried out in the weak topology if that is more convenient. Such a property is exhibited by:

PROPOSITION 8. *If (E, E') is a dual pair and A is a convex subset of E, then $\bar{A}$ is the same for every topology of the dual pair (E, E').*

Proof. We prove that the closure $\bar{A}$ under any topology ξ of (E, E') is the same as the closure $\bar{A}(\sigma)$, say, under $\sigma = \sigma(E, E')$. First, ξ is finer than σ and so $\bar{A} \subseteq \bar{A}(\sigma)$. Next, let $a \notin \bar{A}$. Then (Prop. 5, Cor. 1) there is a continuous linear form $x' \in E'$ with $\langle a, x'\rangle \notin \overline{\langle A, x'\rangle}$. Hence there is some $\delta > 0$ with $|\langle a - x, x'\rangle| \geqslant \delta$ for all $x \in A$. Let $U = \{x : |\langle x, x'\rangle| < \delta\}$. Then U is a neighbourhood in σ such that $a + U$ does not meet A. Thus $a \notin \bar{A}(\sigma)$ and so $\bar{A}(\sigma) \subseteq \bar{A}$.

4. Polar sets. Let (E, E') be a dual pair. If A is a subset of E, the subset of E' consisting of those x' for which

$$\sup\{|\langle x, x'\rangle| : x \in A\} \leqslant 1$$

is called the *polar* of A (in E') and denoted by A^0.

PROPOSITION 9. *Let (E, E') be a dual pair. Then polars in E' of subsets of E have the following properties:*

(i) *A^0 is absolutely convex and $\sigma(E', E)$-closed;*

(ii) *if $A \subseteq B$ then $B^0 \subseteq A^0$;*

(iii) *if $\lambda \neq 0$, then $(\lambda A)^0 = (1/|\lambda|) A^0$;*

(iv) $(\bigcup_\alpha A_\alpha)^0 = \bigcap_\alpha A_\alpha^0$.

Proof. All are immediate from the definition except the $\sigma(E', E)$-closedness of A^0. Now

$$A^0 = \bigcap_{x \in A} \{x' : |\langle x, x'\rangle| \leqslant 1\},$$

which is an intersection of inverse images of closed sets by $\sigma(E', E)$-continuous functions; hence A^0 is $\sigma(E', E)$-closed.

There are some important special cases of polar sets. If M is a vector subspace of E, then $\sup\{|\langle x, x'\rangle| : x \in M\} \leqslant 1$ implies $\langle x, x'\rangle = 0$ for all $x \in M$; hence M^0 consists of those elements of E' that vanish on M and so is the vector subspace of E' *orthogonal* to M. If E is a separated convex space, a subset A' of its dual E' is equicontinuous if and only if there is a neighbourhood U with $|\langle x, x'\rangle| \leqslant 1$ for all $x \in U$ and $x' \in A'$. Thus A' is equicontinuous if and only if it is contained in the polar of some neighbourhood.

Taking the polars of the neighbourhoods in the algebraic dual provides a simple but useful characterisation of the (continuous) dual:

Proposition 10. *If E is a separated convex space and $\mathscr{U}$ is a base of neighbourhoods, then the dual of E is $\bigcup_{U \in \mathscr{U}} U^0$ (the polars being taken in E^*).*

Proof. The linear form $x^* \in E^*$ is continuous if and only if there is some neighbourhood $U \in \mathscr{U}$ with $|\langle x, x^*\rangle| \leqslant 1$ on U.

The operation of taking the polar can be repeated; if (E, E') and (E', F) are dual pairs and A is a subset of E, the polar A^{00} of A^0 in F is called the *bipolar* of A in F. The cases of main interest arise when F is either E or E'^*, but these are both covered by the more general case in which it is supposed only that $E \subseteq F \subseteq E'^*$. This ensures that (E', F) is a dual pair whenever (E, E') is. Then $z \in A^{00}$ if and only if $|\langle z, x'\rangle| \leqslant 1$ whenever $x' \in A^0$, that is, whenever $\sup\{|\langle x, x'\rangle| : x \in A\} \leqslant 1$. Thus $z \in A^{00}$ if and only if

$$|\langle z, x'\rangle| \leqslant \sup\{|\langle x, x'\rangle| : x \in A\}$$

(Ch. I, Lemma 2). Since $A \subseteq E \subseteq F$, this implies $A \subseteq A^{00}$.

Theorem 4. *Let (E, E') be a dual pair and F a vector subspace of E'^* containing E. Then the bipolar A^{00} in F of a subset A of E is the $\sigma(F, E')$-closed absolutely convex envelope of A.*

Proof. Let B be the $\sigma(F, E')$-closed absolutely convex envelope of A. By Proposition 9, A^{00} is a $\sigma(F, E')$-closed absolutely convex set containing A, and therefore $B \subseteq A^{00}$. If $a \notin B$, then (Prop. 5, Cor. 2) there is a continuous linear form $x' \in E'$ with $\langle a, x'\rangle > 1$ and $|\langle x, x'\rangle| \leqslant 1$ for all $x \in B$. Now $A \subseteq B$ and so $x' \in A^0$; thus $a \notin A^{00}$. Hence $A^{00} \subseteq B$ and so $A^{00} = B$.

COROLLARY 1. *If E is a separated convex space with dual E' and A is a subset of E, then the bipolar A^{00} of A in E is the closed absolutely convex envelope of A.*

Proof. By the theorem A^{00} is the $\sigma(E, E')$-closed absolutely convex envelope of A, and, by Proposition 8, $\sigma(E, E')$ may be replaced by the given topology on E.

COROLLARY 2. *Under the conditions of the theorem, the polar of A^{00} in E' is A^0.*

Proof. By the theorem, the polar of A^{00} is the $\sigma(E', F)$-closed absolutely convex envelope of A^0. Now A^0 is absolutely convex and $\sigma(E', E)$-closed; also $\sigma(E', F)$ is finer than $\sigma(E', E)$, because it is a convex topology on E' in which all the linear forms on E' defined by elements of E ($\subseteq F$) are continuous and $\sigma(E', E)$ is the coarsest such topology. Hence A^0 is also $\sigma(E', F)$-closed. Thus A^0 is the polar of A^{00}.

COROLLARY 3. *If (E, E') is a dual pair and if, for each α, A_α is a $\sigma(E, E')$-closed absolutely convex subset of E, then $(\bigcap_\alpha A_\alpha)^0$ is the $\sigma(E', E)$-closed absolutely convex envelope of $\bigcup_\alpha A_\alpha^0$.*

Proof. Taking polars in E of subsets of E',

$$(\bigcup_\alpha A_\alpha^0)^0 = \bigcap_\alpha A_\alpha^{00} = \bigcap_\alpha A_\alpha$$

by Proposition 9 and the theorem; hence

$$(\bigcup_\alpha A_\alpha^0)^{00} = (\bigcap_\alpha A_\alpha)^0,$$

and the result follows by applying the theorem again.

5. Finite-dimensional subspaces. In a separated convex space, these have some especially simple properties, as we shall show in this section.

First consider an n-dimensional vector space E. Corresponding to any base $(e_1, e_2, \ldots, e_n)$ in E, there is a *dual base* $(e_1^*, e_2^*, \ldots, e_n^*)$ in the algebraic dual E^* of E, with the property that $\langle e_i, e_j^* \rangle$ is equal to 1 when $i = j$ and 0 otherwise. For any element x of E can be written uniquely in the form $\sum_{1 \leqslant i \leqslant n} \lambda_i e_i$ and we need only put $\langle x, e_i^* \rangle = \lambda_i$. It is then easy to verify that the e_i^* are linearly independent ($\sum_{1\leqslant j\leqslant n} \mu_j e_j^* = 0$ implies $\mu_i = \langle e_i, \sum_{1\leqslant j\leqslant n} \mu_j e_j^*\rangle = 0$ for each i). They also span E^*, for if $x^* \in E^*$ and $x = \sum_{1\leqslant i\leqslant n} \lambda_i e_i \in E$,

then $$\langle x, x^*\rangle = \sum_{1\leqslant i\leqslant n} \lambda_i \langle e_i, x^*\rangle = \sum_{1\leqslant i\leqslant n} \mu_i \langle x, e_i^*\rangle,$$

with $$\mu_i = \langle e_i, x^*\rangle, \quad \text{and so} \quad x^* = \sum_{1\leqslant i\leqslant n} \mu_i e_i^*.$$

Next, if E is finite-dimensional and (E, E') is a dual pair, then $E' = E^*$. For E and E^* have the same dimension and so have E' and E'^*. Since $E' \subseteq E^*$ and $E \subseteq E'^*$, all must have the same dimension and therefore $E' = E^*$.

Proposition 11. *A finite-dimensional vector space has only one topology under which it is a separated convex space.*

Proof. We show that, for a finite-dimensional separated convex space E, its topology is identical with $\sigma(E, E^*)$. Since the dual of E must be E^*, the topology is certainly finer than $\sigma(E, E^*)$. Now let $(e_1, e_2, \ldots, e_n)$ be any base of E and $(e_1^*, e_2^*, \ldots, e_n^*)$ the corresponding dual base of E^*, and let U be any absolutely convex neighbourhood in E. There is some $\mu > 0$ with $e_i \in \mu U$ for $1 \leqslant i \leqslant n$. Then

$$V = \{x \colon \sup_{1\leqslant i\leqslant n} |\langle x, e_i^*\rangle| \leqslant (\mu n)^{-1}\}$$

is a $\sigma(E, E^*)$-neighbourhood; if $x = \sum_{1\leqslant i\leqslant n} \lambda_i e_i \in V$,

$$x \in \sum_{1\leqslant i\leqslant n} |\lambda_i| \mu U = \sum_{1\leqslant i\leqslant n} |\langle x, e_i^*\rangle| \mu U \subseteq n(\mu n)^{-1} \mu U = U.$$

Thus the given topology is coarser than $\sigma(E, E^*)$, and so identical with it.

This unique topology on a finite-dimensional space E is clearly normable; corresponding to any base $(e_1, e_2, \ldots, e_n)$ we can take, for example, the Euclidean norm

$$\|x\| = \sqrt{\sum_{1\leqslant i\leqslant n} |\lambda_i|^2}$$

for $x = \sum_{1\leqslant i\leqslant n} \lambda_i e_i$. When we have to refer to it, we shall call the unique topology the *Euclidean topology*.

THEOREM 5. *Let M be a finite-dimensional vector subspace of a separated convex space E. Then M is closed in E, and the topology induced on M is the Euclidean topology.*

Proof. The second part is a consequence of Proposition 11, because the induced topology makes M a topological vector space which is locally convex and separated.

For the first part, if $(e_1, e_2, \ldots, e_n)$ is a base of M and if $a \notin M$, then, regarding $a, e_1, e_2, \ldots, e_n$ as linear forms on the dual E' of E, by Lemma 5 there is some $x' \in E'$ with $\langle a, x' \rangle = 1$ and $\langle e_i, x' \rangle = 0$ for $1 \leqslant i \leqslant n$. Let $U = \{x: |\langle x, x' \rangle| < 1\}$. Then U is a neighbourhood (of the origin) and $a + U$ does not meet M. Hence M is closed.

6. The transpose of a linear mapping. Suppose that (E, E') and (F, F') are two dual pairs and that t is a linear mapping of E into F. Then $\langle t(x), y' \rangle$ is a bilinear form in the two variables x and y'. Denote by $t'(y')$ the linear form on E which results from this bilinear form by fixing $y' \in F'$, so that t' is defined by the identity

$$\langle x, t'(y') \rangle = \langle t(x), y' \rangle,$$

valid for all $x \in E$ and all $y' \in F'$. Then for each $y' \in F'$, $t'(y') \in E^*$ and t' is a linear mapping of F' into E^*. We call t' the *transpose* of the linear mapping t (other terminologies include *adjoint*, *conjugate*, and *dual*).

PROPOSITION 12. *Let (E, E') and (F, F') be dual pairs and let t be a linear mapping of E into F with transpose t'. Then $t'(F') \subseteq E'$ if and only if t is continuous when E and F have their weak topologies $\sigma(E, E')$ and $\sigma(F, F')$.*

Proof. First suppose that t is continuous; then for each fixed $y' \in F'$, $\langle t(x), y' \rangle$ is a continuous linear form on E. Hence $t'(y') \in E'$.

Next suppose that $t'(F') \subseteq E'$ and let

$$V = \{y: \sup_{1 \leqslant i \leqslant n} |\langle y, y'_i \rangle| \leqslant 1\}$$

be any $\sigma(F, F')$-neighbourhood. Then if

$$U = \{x: \sup_{1 \leqslant i \leqslant n} |\langle x, t'(y'_i) \rangle| \leqslant 1\},$$

U is a $\sigma(E,E')$-neighbourhood with $t(U) \subseteq V$. Hence t is continuous.

We shall call the linear mapping t *weakly continuous* if it is continuous under the topologies $\sigma(E,E')$ and $\sigma(F,F')$.

The operation of taking the transpose of a linear mapping can clearly be repeated. If (E,E'), (E',E''), (F,F'), (F',F'') are all dual pairs and if t is a weakly continuous mapping of E into F, then t' maps F' into E' and its transpose t'' maps E'' into F'^*. By Proposition 12, t'' maps E'' into F'' if and only if t' is continuous when F' and E' have the topologies $\sigma(F',F'')$ and $\sigma(E',E'')$. The simplest case arises when $E'' = E$ and $F'' = F$; then t'' clearly coincides with t, and so we have, by Proposition 12,

COROLLARY. *If t is weakly continuous, so is its transpose t'.*

We shall have more to say later (Ch. III, Prop. 4) about the continuity of t and t' under other topologies.

PROPOSITION 13. *If t is a continuous linear mapping of the separated convex space E (with dual E') into the separated convex space F (with dual F'), then t is also continuous when E and F have the associated weak topologies $\sigma(E,E')$ and $\sigma(F,F')$.*

Proof. For each fixed $y' \in F'$, $\langle t(x), y'\rangle$ is a continuous linear form on E and so $t'(y') \in E'$. Hence $t'(F') \subseteq E'$ and the result follows from Proposition 12.

The converse of this result is clearly not valid in general (take, for example, the identity mapping of E under one topology into E under a strictly finer one of the same dual pair). But we shall see (Ch. III, Prop. 14) that for a suitable topology on E, weak continuity implies continuity in the initial topologies.

There is an algebraic relation which we shall require frequently in later chapters.

LEMMA 6. *Let (E,E') and (F,F') be dual pairs and let t be a weakly continuous linear mapping of E into F, with transpose t'. Then, for each subset A of E,*

$$(t(A))^0 = t'^{-1}(A^0).$$

Proof. Each of these sets is the set of all $y' \in F'$ with

$$|\langle t(x), y'\rangle| = |\langle x, t'(y')\rangle| \leqslant 1$$

for all $x \in A$.

SUPPLEMENT

In this section we describe the duals of some of the spaces given in the Supplement to Chapter I.

(1) *Finite dimensional spaces.* See § 5.

(2) *Spaces of continuous functions.* The Riesz representation theorem (see e.g. S. Banach, *Théorie des opérations linéaires* (1932), Ch. IV, § 4) states that any continuous linear form f on the space $\mathscr{C}(I)$ of functions continuous on the interval $I = [a, b]$ can be expressed as a Stieltjes integral

$$f(x) = \int_a^b x(t)\, dg(t),$$

where g is a function of bounded variation. Thus the dual of $\mathscr{C}(I)$ can be identified with the space of Stieltjes measures on $[a, b]$.

This result encourages us to define a measure μ on a compact set S to be a continuous linear form on the normed space $\mathscr{C}(S)$ (Ch. I, Suppl. 2 a), and to write

$$\mu(x) = \int_S x(t)\, d\mu.$$

When S is not compact, we cannot hope to get a useful definition of a measure as an element of the dual of $\mathscr{C}(S)$, because, for example, such a definition would exclude ordinary Lebesgue measure on $]-\infty, \infty[$ (for $\int_{-\infty}^{\infty} x(t)\, dt$ certainly does not exist for all functions x continuous on $]-\infty, \infty[$). If S is locally compact and separated, we are led to define a (*Radon*) *measure* μ on S to be a linear form on $\mathscr{K}(S)$ which is continuous on $\mathscr{K}_A(S)$ for each compact set A (see Ch. I, Suppl. 2 c). We shall prove (Ch. V, Prop. 5) that this is equivalent to the requirement that μ be continuous under the inductive limit topology on $\mathscr{K}(S)$. For a systematic development of integration theory on these lines see N. Bourbaki, *Eléments de mathématique, Intégration* (1952).

Under the coarser topology of uniform convergence, the dual of $\mathscr{K}(S)$ is a subset of the set of measures, namely the set of those measures that are restrictions to S of measures on the one-point compactification of S. Finally, under the still coarser topology of compact convergence, the space $\mathscr{K}(S)$ is dense in $\mathscr{C}(S)$ and the dual of both is the set of measures of compact support (the *support* of the measure μ is the smallest closed set A such that $\mu(x) = 0$ for all x whose support does not meet A).

(3) *Spaces of indefinitely differentiable functions.* The continuous linear forms on the space $\mathscr{D}$ of real (or complex) valued indefinitely differentiable functions of compact support, under the inductive limit topology (Ch. I, Suppl. 3 *d*) are called *distributions.* Now $\mathscr{D}$ can be considered as a vector subspace of $\mathscr{K}(]-\infty, \infty[)$, which (under its inductive limit topology) induces a coarser topology on $\mathscr{D}$. Hence every measure is a distribution (but, for example, the distribution f defined by $f(x) = x'(0)$ is not a measure).

Since the topology of $\mathscr{D}$ is finer that that induced on it by the topology of $\mathscr{S}$ (Ch. I, Suppl. 3 *c*), any continuous linear form on $\mathscr{S}$ is a distribution, said to be of 'slow increase' (see L. Schwartz, *Théorie des distributions*, II (1951), Ch. VII, § 4).

The dual $\mathscr{E}'$ of the space $\mathscr{E}$ of indefinitely differentiable functions under the topology of compact convergence for all derivatives (Ch. I, Suppl. 3 *b*) is, similarly, a subspace of the space $\mathscr{D}'$ of distributions; the elements of $\mathscr{E}'$ are distributions of compact support (the support of a distribution being defined in the same way as that of a measure).

On the space $\mathscr{D}$, the operation of differentiation is an example of a continuous linear mapping of the space into itself. Differentiation of distributions is defined by $\dot{f}(x) = -f(\dot{x})$, where dots denote differentiation; thus the operation of differentiation on $\mathscr{D}'$ is the negative of the transpose of the operation of differentiation on $\mathscr{D}$.

(4) *Spaces of holomorphic functions.* Any continuous linear form on $\mathscr{H}(D)$ can be expressed in the form

$$f(x) = (2\pi i)^{-1} \int_\Gamma x(t)\, \phi(t)\, dt,$$

where ϕ is holomorphic in a domain D' containing $\sim D$, $\phi(\infty) = 0$ and Γ is a closed contour in $D \cap D'$. Also any such linear form is continuous. Different functions ϕ give the same linear form if they agree on a domain containing $\sim D$. This latter is an equivalence relation among functions holomorphic on domains containing $\sim D$. The equivalence classes are called locally holomorphic functions on $\sim D$; they form a vector space which is the dual of $\mathscr{H}(D)$. (See G. Köthe, 'Dualität in der Funktionentheorie', *J. Reine Angew. Math.* 191 (1953), 30–50.)

(5) *Sequence spaces.* Dual pairs of spaces whose elements are sequences of scalars can be formed in the following way. Let A be any set of sequences $z = (z_n)$, and let E' be the vector space of all sequences $x' = (x'_n)$ such that $\sum_{n=1}^{\infty} |z_n x'_n| < \infty$ for all $z \in A$. Then let E be the vector space of all sequences $x = (x_n)$ such that $\sum_{n=1}^{\infty} |x_n x'_n| < \infty$ for all $x' \in E'$. (Clearly $A \subseteq E$.) The bilinear form $\langle x, x' \rangle = \sum_{n=1}^{\infty} x_n x'_n$ makes (E, E') a dual pair. Then E (and E') can be topologised in any of the general ways discussed in Chapter III. (See G. Köthe, 'Die Stufenräume, eine einfache Klasse linearer vollkommener Räume', *Math. Z.* 51 (1948), 317–45.)

If A consists of the sequence $z = (0, 0, \ldots)$ alone, E' is the space of all sequences and E the space of sequences with at most a finite number of non-zero terms. In this case E' is also the algebraic dual E^* of E. If A consists of the sequence $z = (1, 1, \ldots)$ alone, E' is the space l^1 of absolutely convergent sequences and E the space m of bounded sequences.

The dual of the normed spaces c and c_0 (see Ch. I, Suppl. 5) is l^1, the dual of l^1 is m $(= l^\infty)$, and the dual of l^p $(1 < p < \infty)$ is l^q, where $p^{-1} + q^{-1} = 1$. The dual of the metric space l^p for $0 < p < 1$ is m, an example of a space which is not locally convex but has an adequate dual, (l^p, m) being a dual pair. The dual of the normed space m is not a sequence space at all; in fact it is the space of bounded finitely additive set functions defined on the subsets of the set of positive integers.

(6) *Integration spaces.* There is a duality theory of these spaces which parallels the theory for sequence spaces, using this time the bilinear form $\langle x, x' \rangle = \int_S x(t)\, x'(t)\, d\mu$. (See J. Dieudonné, 'Sur les espaces de Köthe', *J. Analyse Math.* 1 (1951), 81–115.)

The dual of the normed space $\mathscr{L}^p$ $(1 < p < \infty)$ is $\mathscr{L}^q$ (where $p^{-1}+q^{-1} = 1$); in most cases (e.g. when S has finite or σ-finite measure) the dual of $\mathscr{L}^1$ is $\mathscr{M}(=\mathscr{L}^\infty)$. But if $0 < p < 1$ the dual of the metric space $\mathscr{L}^p(0,1)$ is $\{o\}$, an example of a non-locally convex space which has a hopelessly inadequate dual. (If f is a non-zero linear form, there is an element $x \in \mathscr{L}^p(0,1)$ with $d(x,o) = 1$ and $f(x) = \alpha > 0$. Then if $\phi(s) = \int_0^s |x(t)|^p\,dt$, ϕ is a continuous function of s and so $[0,1]$ can be dissected by points $0 = s_0 < s_1 < \dots < s_n = 1$ so that $\phi(s_r)-\phi(s_{r-1}) = n^{-1}$. Put $y_r(t) = x(t)$ in $[s_{r-1}, s_r]$ and zero otherwise. Then

$$x = y_1+y_2+\dots+y_n$$

and so, for at least one r, $|f(y_r)| \geqslant n^{-1}\alpha$. Let $x_n = ny_r$ for this r. Then $|f(x_n)| \geqslant \alpha$ but $d(x_n, o) = n^{p-1} \to 0$. Hence f cannot be continuous.)

(7) *Finest convex topology.* The dual of a vector space E under its finest convex topology is E^*. (For, if $x^* \in E^*$, $\{x\colon |\langle x, x^*\rangle| \leqslant 1\}$ is a neighbourhood.)

(8) *Hilbert space.* Any continuous linear form f on a Hilbert space E can be expressed in terms of the inner product by the formula $f(x) = (x, a)$ for some element a of E. Writing $a = \theta(f)$, we have a conjugate-isomorphism θ of the dual E' of E onto E (θ is $(1,1)$ and onto and $\theta(f+g) = \theta(f)+\theta(g)$, but $\theta(\lambda f) = \bar{\lambda}\theta(f)$).

CHAPTER III

TOPOLOGIES ON DUAL SPACES AND THE MACKEY–ARENS THEOREM

We continue with the study of duality in this chapter. First we consider a general method of defining convex topologies on the dual of a convex space, taking as neighbourhoods of the origin the polars of certain sets in the convex space. The sets that are suitable form the important class of bounded sets. The main result is the Mackey–Arens theorem, which characterises all convex topologies with a given dual. It turns out, most elegantly, that all such topologies lie between a coarsest (the weak topology) and a finest, which we obtain. This chapter makes heavier demands on general topology than the previous two; compact and precompact subsets of a convex space enter into the development here and we use the notion of a complete space. But we develop the necessary topology as we go along.

1. Bounded sets. In a vector space the set A is said to *absorb* the set B if there is some $\alpha > 0$ such that $B \subseteq \lambda A$ for all λ with $|\lambda| \geqslant \alpha$. An absorbent set A is one which absorbs points.

In a convex space E, a set A is called *bounded* if it is absorbed by every neighbourhood (of the origin). If $\mathscr{U}$ is a base of absolutely convex neighbourhoods, the set A is bounded if and only if to each $U \in \mathscr{U}$ corresponds a positive λ with $A \subseteq \lambda U$ (this implies $A \subseteq \mu U$ for all $|\mu| \geqslant \lambda$). If the topology of E is determined by the set Q of seminorms, so that the sets

$$\{x\colon \sup_{1 \leqslant i \leqslant n} p_i(x) \leqslant \epsilon\} \quad (p_i \in Q)$$

form a base of neighbourhoods, then A is bounded if and only if $p(A)$ is a bounded set of real numbers for each $p \in Q$. In particular A is bounded in $\sigma(E, E')$ if and only if $\langle A, x'\rangle$ is bounded for each $x' \in E'$.

Lemma 1. (i) *The closure, convex envelope and absolutely convex envelope of a bounded set are bounded;*

(ii) *any subset or any scalar multiple of a bounded set is bounded;*
(iii) *any finite union or sum of bounded sets is bounded.*

Proof. (i) Take a base of closed absolutely convex neighbourhoods U. If A is bounded and B any of the sets in (i), then $A \subseteq \lambda U$ implies $B \subseteq \lambda U$. Parts (ii) and (iii) are easy consequences of the definition.

PROPOSITION 1. *The image by a continuous linear mapping of a bounded set is bounded.*

Proof. Suppose that t is a continuous linear mapping of E into F and that A is a bounded subset of E. For each absolutely convex neighbourhood V in F, $t^{-1}(V)$ is a neighbourhood in E and so there is some λ with $A \subseteq \lambda t^{-1}(V)$. Thus $t(A) \subseteq \lambda V$; hence $t(A)$ is bounded in F.

In a normed space the balls $\{x: \|x\| \leqslant \epsilon\}$ form a neighbourhood base consisting of bounded sets. This property characterises normed spaces:

THEOREM 1. *If a separated convex space contains a bounded neighbourhood, it is normable.*

Proof. Let U be an absolutely convex neighbourhood contained in the bounded neighbourhood; then U is bounded (Lemma 1 (ii)). Hence, if V is any neighbourhood, there is some $\lambda > 0$ with $U \subseteq \lambda V$ and so $(1/\lambda)\, U \subseteq V$. Thus the sets $\{\epsilon U\}$ $(\epsilon > 0)$ form a neighbourhood base. Since the space is separated, the gauge of U is a norm defining the topology.

The theorem shows incidentally that the balls $\{x: d(x, o) \leqslant \epsilon\}$ of a metrisable non-normable convex space are not bounded sets.

If ξ and η are two convex topologies on E with ξ finer than η, a set bounded in ξ remains bounded in η. In particular, if E' is the dual of E under the separated topology ξ, then the ξ-bounded sets are also weakly bounded (i.e. $\sigma(E, E')$-bounded). We shall see in Theorem 1 of Chapter IV that the converse is true; the same sets are bounded in all topologies of the dual pair (E, E').

LEMMA 2. *For a dual pair (E, E') the following three assertions are equivalent:*

(i) *A is a weakly bounded subset of E;*
(ii) *$p'(x') = \sup\{|\langle x, x'\rangle| : x \in A\}$ is a seminorm on E';*
(iii) *A^0 is an absorbent subset of E'.*

Proofs are immediate from the definitions.

Thus the polars of weakly bounded sets, being also absolutely convex, have all the properties required of neighbourhoods. This suggests the following way of topologising E'.

2. Polar topologies. Let (E, E') be a dual pair and $\mathscr{A}$ any set of weakly bounded subsets of E. Then the sets $A^0 (A \in \mathscr{A})$ are absolutely convex and absorbent, and so (Ch. I, Th. 2, Corollary) there is a coarsest topology ξ' on E' in which they are neighbourhoods. A base of neighbourhoods in ξ' is formed by the sets

$$\epsilon \bigcap_{1 \leqslant i \leqslant n} A_i^0 = (\epsilon^{-1} \bigcup_{1 \leqslant i \leqslant n} A_i)^0 \quad (\epsilon > 0, A_i \in \mathscr{A}).$$

It turns out that convergence in ξ' means uniform convergence on each $A \in \mathscr{A}$, and so the topology ξ' is called the *topology of uniform convergence on the sets of* $\mathscr{A}$, or the *topology of* $\mathscr{A}$*-convergence*. We shall call any topology defined in this way a *polar topology*.

In practice it nearly always happens that $\mathscr{A}$ satisfies:

B 1: *if* $A \in \mathscr{A}$ *and* $B \in \mathscr{A}$ *then there is some* $C \in \mathscr{A}$ *with* $A \cup B \subseteq C$;

B 2: *if* $A \in \mathscr{A}$ *and* λ *is any scalar, then* $\lambda A \in \mathscr{A}$;

B 3: $\bigcup_{A \in \mathscr{A}} A$ *spans* E.

The conditions B 1 and B 2 ensure that the polars of the sets of $\mathscr{A}$ form a neighbourhood base for ξ', because then if $\epsilon > 0$ and $A_i \in \mathscr{A}$ for $1 \leqslant i \leqslant n$, there is some $C \in \mathscr{A}$ with

$$\epsilon^{-1} \bigcup_{1 \leqslant i \leqslant n} A_i \subseteq C \quad \text{and so} \quad C^0 \subseteq \epsilon \bigcap_{1 \leqslant i \leqslant n} A_i^0.$$

Also B 3 is a sufficient condition for each $x \in E$ to be bounded on some A^0 and so to define a ξ'-continuous linear form on E'. Thus ξ' is finer than $\sigma(E', E)$ and so is separated.

For convenience, we shall almost always assume that when a topology of $\mathscr{A}$-convergence is referred to, the conditions B 1, B 2 and B 3 are satisfied. (Conditions B 1 and B 2 are not a serious restriction, for if $\mathscr{A}$ is any set of weakly bounded subsets of E, the set of all scalar multiples of finite unions of sets of $\mathscr{A}$ satisfies B 1 and B 2 and clearly defines the same polar topology.) Another assumption, which is often convenient and which we shall sometimes make explicitly, is that the sets of $\mathscr{A}$ are weakly closed and absolutely convex. This again involves no loss of generality. For if $\mathscr{A}$ satisfies B 1, B 2 and B 3, the set of closed absolutely convex envelopes of the sets of $\mathscr{A}$ also satisfies B 1, B 2 and B 3; since they have the same polars, they define the same polar topology.

Alternatively, a polar topology can be described by seminorms. For each $A \in \mathscr{A}$, put

$$p'_A(x') = \sup\{|\langle x, x'\rangle| : x \in A\};$$

then p'_A is the gauge of A^0 and the set $\{p'_A : A \in \mathscr{A}\}$ determines the topology of $\mathscr{A}$-convergence.

When $\mathscr{A}$ is the set of all finite subsets of E, the topology of $\mathscr{A}$-convergence is $\sigma(E', E)$, which is the coarsest polar topology (because of B 3). The finest polar topology is obtained by taking $\mathscr{A}$ to be the set of all weakly bounded subsets of E. This topology is denoted by $\beta(E', E)$ and is sometimes called the *strong topology* on E'. We shall come across some intermediate topologies in § 4.

If E is a normed space and $\mathscr{A}$ is taken to be the set of all balls $\{x: \|x\| \leqslant \epsilon\}$ $(\epsilon > 0)$, the topology of $\mathscr{A}$-convergence is normable, with norm $\|x'\| = \sup\{|\langle x, x'\rangle| : \|x\| \leqslant 1\}$. When we have shown in Chapter IV that the norm-bounded sets of E are the same as the weakly bounded sets, we shall have proved that this norm topology on E' is $\beta(E', E)$.

The dual of E' under the coarsest polar topology is E; in general the dual may be a larger subspace of E'^*:

PROPOSITION 2. *If (E, E') is a dual pair, the dual of E' under the topology of $\mathscr{A}$-convergence is* $\bigcup_{A \in \mathscr{A}} A^{00}$, *the bipolars being taken in E'^*.*

Proof. This follows from Proposition 10 of Chapter II.

The natural problem of finding the condition on $\mathscr{A}$ for the dual of E' under the topology of $\mathscr{A}$-convergence to be exactly E requires more topology than we have at our disposal at present; it is solved in § 7.

The rôles of E and E' can clearly be interchanged and E given polar topologies. In fact:

PROPOSITION 3. *Every separated convex topology is a polar topology, namely, the topology of uniform convergence on the equicontinuous subsets of the dual space.*

Proof. Let $\mathscr{U}$ be a base of closed absolutely convex neighbourhoods for the separated convex topology ξ. Then $U = U^{00}$ (Ch. II, Th. 4, Cor. 1) and so ξ is the topology of uniform convergence on the sets U^0 ($U \in \mathscr{U}$). If A' is equicontinuous, there is some $U \in \mathscr{U}$ with $A' \subseteq U^0$; thus ξ is the topology of uniform convergence on the equicontinuous subsets of the dual.

Now that we can topologise both E and E' in a general way, we can improve upon an earlier result (Ch. II, Prop. 12, Corollary), which stated that if t is a weakly continuous mapping then so is its transpose t'.

PROPOSITION 4. *Suppose that (E, E') and (F, F') are dual pairs and that t is a weakly continuous mapping of E into F. If E' has the topology of uniform convergence on the sets of $\mathscr{A}$ and if F' has the topology of uniform convergence on the sets of $t(\mathscr{A})$, then t' is continuous.*

Proof. If $A \in \mathscr{A}$, then $(t(A))^0 = t'^{-1}(A^0)$ (Ch. II, Lemma 6) and so t' is continuous.

COROLLARY. *If E and F are normed spaces and if E' and F' have their norm topologies, then t is continuous if and only if t' is continuous.*

Proof. Let $\mathscr{A}$ be the set of balls $\{x: \|x\| \leqslant \epsilon\}$ in E and similarly $\mathscr{B}$ the set of balls in F. If t is continuous and if $A \in \mathscr{A}$, then there is some $B \in \mathscr{B}$ with $t(A) \subseteq B$, and so $B^0 \subseteq (t(A))^0$. Hence the topology of uniform convergence on the sets of $t(\mathscr{A})$ is coarser than the norm topology on F'. But t' is continuous, by the proposition, when F' has the topology of uniform convergence on the sets of $t(\mathscr{A})$ and so certainly when F' has the norm topology. The same argument, applied with t' for t and with F' and E' for E and F, shows that if t' is continuous so is t.

3. Precompact sets. In a metrisable convex space the 'size' of a subset A can be measured by its diameter

$$d(A) = \sup\{d(x, y): x, y \in A\}.$$

This can be expressed in terms of neighbourhoods by putting $U_\epsilon = \{x: d(x, o) \leqslant \epsilon\}$. Then $d(A) \leqslant \epsilon$ if and only if $x - y \in U_\epsilon$ for all $x, y \in A$. In this form, the idea generalises to any convex space. If U is any neighbourhood in a convex space, the subset A is called *small of order* U if $x - y \in U$ for all $x, y \in A$.

A subset A of a convex space is called *precompact* if, for every (absolutely convex) neighbourhood U, A can be covered by a finite number of sets $A_1, A_2, \ldots, A_n$ small of order U. If $a_i \in A_i$ for each i then

$$A \subseteq \bigcup_{1 \leqslant i \leqslant n} (a_i + U).$$

Conversely, a set A with the property that for each neighbourhood U there are points $a_1, a_2, \ldots, a_n$ with

$$A \subseteq \bigcup_{1 \leqslant i \leqslant n} (a_i + U)$$

is precompact. Tapioca would make a suitable mental image.

Lemma 3. (i) *The closure of a precompact set is precompact;*

(ii) *any subset or any scalar multiple of a precompact set is precompact;*

(iii) *any finite union or sum of precompact sets is precompact.*

Proof. (i) For a closed absolutely convex neighbourhood U,

$$A \subseteq \bigcup_{1 \leqslant i \leqslant n} (a_i + U) \quad \text{implies} \quad \bar{A} \subseteq \bigcup_{1 \leqslant i \leqslant n} (a_i + U),$$

the union being closed. Parts (ii) and (iii) are direct from the definition; in (iii), if U is absolutely convex and if

$$A \subseteq \bigcup_{1 \leqslant i \leqslant n} (a_i + U) \quad \text{and} \quad B \subseteq \bigcup_{1 \leqslant j \leqslant m} (b_j + U),$$

then
$$A + B \subseteq \bigcup_{\substack{1 \leqslant i \leqslant n \\ 1 \leqslant j \leqslant m}} (a_i + b_j + 2U).$$

A precompact set remains precompact in any coarser topology; the image of a precompact set by a continuous linear mapping is precompact.

Proposition 5. *A precompact set is bounded.*

Proof. If U is an absolutely convex neighbourhood and

$$A \subseteq \bigcup_{1 \leqslant i \leqslant n} (a_i + U),$$

there is some $\lambda > 0$ with $a_i \in \lambda U$ for $1 \leqslant i \leqslant n$. Then $A \subseteq (1 + \lambda) U$.

The converse is true only in special cases (see Proposition 6). Because precompactness is in general a stronger condition than boundedness, a precompact set is sometimes called *totally bounded*. In a normed space there are neighbourhoods which are bounded, but they are unlikely to be precompact:

THEOREM 2. *If a separated convex space has a precompact neighbourhood, it is finite-dimensional.*

Proof. By Proposition 5, the space E has a bounded neighbourhood and so is normable, by Theorem 1. Let U be the unit ball $\{x: \|x\| \leqslant 1\}$; the given precompact neighbourhood contains λU for some $\lambda > 0$ and thus U is precompact, by Lemma 3. There are points $a_1, a_2, \ldots, a_n$ with $U \subseteq \bigcup_{1 \leqslant i \leqslant n} (a_i + \frac{1}{2}U)$. Let M be the vector subspace spanned by $a_1, a_2, \ldots, a_n$. Then

$$U \subseteq M + \tfrac{1}{2}U \subseteq M + \tfrac{1}{2}(M + \tfrac{1}{2}U) = M + (\tfrac{1}{2})^2\, U \subseteq \ldots \subseteq M + (\tfrac{1}{2})^k\, U$$

for any integer k. Hence $U \subseteq \overline{M}$. But $\overline{M} = M$ since M is finite-dimensional (Ch. II, Th. 5) and

$$E = \bigcup_{\lambda > 0} \lambda U \subseteq M.$$

LEMMA 4. *Suppose that $\mathscr{V}$ is a set of absolutely convex neighbourhoods in the convex space E, such that the finite intersections of the sets of $\mathscr{V}$ form a base of neighbourhoods. If the set A has a finite covering by sets small of order V for each $V \in \mathscr{V}$ then A is precompact.*

Proof. First, if V, $W \in \mathscr{V}$ and $U = V \cap W$ then there are sets $B_1, B_2, \ldots, B_m$ small of order V with

$$A \subseteq \bigcup_{1 \leqslant r \leqslant m} B_r$$

and sets $C_1, C_2, \ldots, C_n$ small of order W with

$$A \subseteq \bigcup_{1 \leqslant s \leqslant n} C_s.$$

Then the sets $B_r \cap C_s$ $(1 \leqslant r \leqslant m,\ 1 \leqslant s \leqslant n)$ form a finite covering of A by sets small of order U.

The general case when U is an intersection of k sets of $\mathscr{V}$ can now be treated by induction on k.

PROPOSITION 6. *Let (E, E') be a dual pair and A a weakly bounded subset of E. Then A is weakly precompact.*

Proof. In Lemma 4 take for $\mathscr{V}$ the sets $\{x: |\langle x, x'\rangle| \leqslant 1\}$ $(x' \in E')$. For each $x' \in E'$, $\langle A, x'\rangle$ is a bounded set of scalars and so can be covered by a finite number of sets $\Gamma_1, \Gamma_2, \ldots, \Gamma_n$, each of diameter less than 1. Then the sets $x'^{-1}(\Gamma_i)$ form a finite covering of A by sets small of order V, where $V = \{x: |\langle x, x'\rangle| \leqslant 1\}$.

THEOREM 3. *Let (E, E') and (F, F') be dual pairs and t a weakly continuous linear mapping of E into F with transpose t'. Let $\mathscr{A}$ and $\mathscr{B}'$ be sets of weakly bounded subsets of E and F' defining topologies of uniform convergence on E' and F. Then the following are equivalent:*

(i) *for each $A \in \mathscr{A}$, $t(A)$ is precompact under the topology of $\mathscr{B}'$-convergence,*

(ii) *for each $B' \in \mathscr{B}'$, $t'(B')$ is precompact under the topology of $\mathscr{A}$-convergence.*

Proof. Assume (ii). If $A \in \mathscr{A}$ and $B' \in \mathscr{B}'$ then there is a finite set K' with $t'(B') \subseteq K' + \frac{1}{3}A^0$. Since A is weakly bounded, it is weakly precompact by Proposition 6, and so there is a finite covering of A by sets A_i small of order $\frac{1}{3}K'^0$. If x and y are points of $A \cap A_i$ and $b' \in B'$, then $t'(b') = c' + a'$ ($c' \in K'$ and $a' \in \frac{1}{3}A^0$), and

$$\begin{aligned} |\langle t(x-y), b'\rangle| &= |\langle x-y, t'(b')\rangle| \\ &\leqslant |\langle x-y, c'\rangle| + |\langle x, a'\rangle| + |\langle y, a'\rangle| \\ &\leqslant \tfrac{1}{3} + \tfrac{1}{3} + \tfrac{1}{3}. \end{aligned}$$

Thus the sets $t(A \cap A_i)$ form a finite covering of $t(A)$ and are small of order B'^0.

COROLLARY 1. *Let (E, E') be a dual pair. Then each $A \in \mathscr{A}$ is precompact in the topology of $\mathscr{A}'$-convergence if and only if each $A' \in \mathscr{A}'$ is precompact in the topology of $\mathscr{A}$-convergence.*

COROLLARY 2. *The convex envelope, absolutely convex envelope and closed absolutely convex envelope of a precompact subset of a separated convex space are precompact.*

Proof. Let $\mathscr{A}$ be the set of precompact subsets of the space E and $\mathscr{B}$ the set of their absolutely convex envelopes. Let $\mathscr{A}'$ be the set of equicontinuous subsets of the dual. Then each $A \in \mathscr{A}$ is precompact in the topology of $\mathscr{A}'$-convergence and so by Corollary 1 each $A' \in \mathscr{A}'$ is precompact in the topology of $\mathscr{A}$-convergence and therefore in the topology of $\mathscr{B}$-convergence, since these topologies coincide. Hence, again by Corollary 1, each $B \in \mathscr{B}$ is precompact in the topology of $\mathscr{A}'$-convergence, that is, a precompact subset of E. The two other cases follow from Lemma 3 (i) and (ii).

4. Compact sets. In many respects the part played by bounded sets in a finite-dimensional space is taken over not so much by bounded sets in a convex space but rather by the smaller class of compact sets. We give the definition for a general topological space.

Suppose that E is a topological space and that A is a subset of E. A set $\mathscr{C}$ of subsets of E is said to form a *covering* of A if A is contained in the union of the sets of $\mathscr{C}$. If also all the sets of $\mathscr{C}$ are open, $\mathscr{C}$ is called an *open covering* of A. The set A is called *compact* if every open covering of A contains a finite subcovering. Clearly every finite set is compact; so also is any set formed by a convergent sequence together with its limit. (For any open set which contains the limit contains all but a finite number of points of the sequence.)

In the topology induced on any subset A of E, the open sets are the intersections with A of the open sets of E. Hence A is a compact subset of E if and only if it is a compact subset of itself under the induced topology. Thus compactness is an 'intrinsic' property of a set, unlike, for example, the property of being closed. This is a special case of:

LEMMA 5. *Suppose that ξ and η are two topologies on the same space and that ξ is finer than η on the set A. If A is ξ-compact then A is also η-compact.*

Proof. This follows easily from the fact that a set open in the topology induced on A by η is also open in the topology induced by ξ.

LEMMA 6. *In a topological space:*

(i) *any finite union of compact sets is compact;*

(ii) *any closed subset of a compact set is compact;*

(iii) *any compact subset of a separated space is closed.*

Proofs. The first part is immediate. (ii) If B is a closed subset of the compact set A and $\mathscr{C}$ is an open covering of B, then $\mathscr{C} \cup \{\sim B\}$ is an open covering of A. This contains a finite subcovering of A which possibly includes $\sim B$; the remaining sets form a finite subcovering of B. (iii) Let A be compact and $a \notin A$. Then for each $x \in A$ there are disjoint open sets C_x and B_x with $x \in C_x$ and $a \in B_x$, since the space is separated. The sets C_x form an open covering of A and so there is a finite subcovering

$$\{C_{x_i}: 1 \leqslant i \leqslant n\}.$$

Then $\bigcap_{1 \leqslant i \leqslant n} B_{x_i}$ is a neighbourhood of a not meeting A. Hence $a \notin \bar{A}$. Thus A is closed.

PROPOSITION 7. *The image by a continuous mapping of a compact set is compact.*

Proof. Let A be compact. The inverse image by a continuous mapping f of an open covering of $f(A)$ is an open covering of A and so contains a finite subcovering; the corresponding sets in the original open covering of $f(A)$ form the required subcovering.

COROLLARY. *A continuous real-valued function is bounded and attains its bounds on every compact set.*

For if A is compact and f continuous, $f(A)$ is a compact set of real numbers. This is closed, by Lemma 6 (iii). It is also bounded, because each point of $f(A)$ is the centre of an open interval of unit length and a finite number of them covers $f(A)$. Hence f is bounded on A; if $\alpha = \sup f(A)$, then $\alpha \in \overline{f(A)} = f(A)$ and so is attained. Similarly $\inf f(A)$ is attained.

Suppose now that E is a convex space. It follows directly from the definitions that a compact subset of E is precompact and thus, by Proposition 5, bounded. For if A is compact and U an open neighbourhood, then $\{x + U: x \in A\}$ is an open covering of A and so there is a finite number of points $x_1, x_2, \ldots, x_n$ with $A \subseteq \bigcup_{1 \leqslant i \leqslant n} (x_i + U)$.

LEMMA 7. *In a convex space:*

(i) *any scalar multiple of a compact set is compact;*

(ii) *any finite sum of compact sets is compact;*

(iii) *the sum of a compact set and a closed set is closed.*

Proofs. The first part is easy. (ii) Let A and B be compact and $\mathscr{C}$ an open covering of $A+B$. For each $x \in A$ and each $y \in B$, there is an open absolutely convex neighbourhood $U(x, y)$ of the origin for which $x+y+U(x, y)$ is contained in some set of $\mathscr{C}$. If x is fixed, the sets $y+\frac{1}{2}U(x, y)$ form an open covering of B; let

$$\{y_j+\tfrac{1}{2}U(x, y_j)\colon 1 \leqslant j \leqslant n(x)\}$$

be a finite subcovering and put

$$V(x) = \bigcap_{1 \leqslant j \leqslant n(x)} \tfrac{1}{2}U(x, y_j).$$

Then the sets $x+V(x)$ form an open covering of A; let

$$\{x_i+V(x_i)\colon 1 \leqslant i \leqslant m\}$$

be a finite subcovering. Then

$$\begin{aligned} A+B &\subseteq \bigcup_{1 \leqslant i \leqslant m} (x_i+V(x_i))+B \\ &\subseteq \bigcup_{1 \leqslant i \leqslant m} \bigcup_{1 \leqslant j \leqslant n(x_i)} (x_i+y_j+\tfrac{1}{2}U(x_i, y_j)+\tfrac{1}{2}U(x_i, y_j)) \end{aligned}$$

and this last set is contained in a finite union of sets of $\mathscr{C}$. Thus $A+B$ is compact. (iii) Let A be compact and B closed and $a \notin A+B$. Then for each $x \in A$, $x+B$ is closed and so there is an open absolutely convex neighbourhood $U(x)$ of the origin with $a+U(x)$ not meeting $x+B$. Then $a \notin x+U(x)+B$. Now the sets $x+\frac{1}{2}U(x)$ form an open covering of the compact set A; let $\{x_i+\frac{1}{2}U(x_i)\colon 1 \leqslant i \leqslant n\}$ be a finite subcovering and put

$$V = \bigcap_{1 \leqslant i \leqslant n} \tfrac{1}{2}U(x_i).$$

Then

$$A+V \subseteq \bigcup_{1 \leqslant i \leqslant n} (x_i+\tfrac{1}{2}U(x_i)+\tfrac{1}{2}U(x_i)) \subseteq \bigcup_{x \in A} (x+U(x))$$

and so $a \notin A+V+B$. Thus $a+V$ does not meet $A+B$ and so $a \notin \overline{A+B}$. Hence $A+B$ is closed.

Suppose now that E is a separated convex space. The set of compact subsets of E can be used to define on the dual E' a polar topology which is intermediate between $\sigma(E', E)$ and $\beta(E', E)$. Now if A is bounded (or precompact), the absolutely convex envelope of A is also bounded (or precompact), by Lemma 1 and Corollary 2 of Theorem 3. It is, however, not true in general that the closed absolutely convex envelope of a compact set is compact (but for an important special case when this is true see § 6, Corollary of Th. 5; see also Ch. VI, Suppl. 4). It turns out that an especially important role is played by the topology on E' of uniform convergence on the absolutely convex compact subsets of E. These sets satisfy the conditions B 1, B 2 and B 3. For B 2 and B 3 are obvious; to show that B 1 is fulfilled, let A and B be absolutely convex compact subsets of E. Then the closed absolutely convex envelope of $A \cup B$ is a subset of $A+B$ and so is compact, by Lemma 6 (ii) and Lemma 7. We can now identify the dual of E' under the topology of uniform convergence on the absolutely convex compact subsets of E. First we have:

LEMMA 8. *If A is an absolutely convex compact subset of the separated convex space E then $A^{00} = A$ (the bipolar being taken in E'^*).*

Proof. By Theorem 4 of Chapter II, A^{00} is the $\sigma(E'^*, E')$-closure of A. Now A is compact in the topology ξ, say, of E; $\sigma(E'^*, E')$ induces on E the topology $\sigma(E, E')$, which is coarser than ξ, and so A is also $\sigma(E'^*, E')$-compact (Lemma 5). Hence A is $\sigma(E'^*, E')$-closed (Lemma 6 (iii)) and so $A^{00} = A$.

PROPOSITION 8. *Let E be a separated convex space. Then the dual of E' is E when E' has the topology of uniform convergence on a set of absolutely convex compact subsets of E.*

Proof. The dual of E' is $\bigcup_{A \in \mathscr{A}} A^{00}$ (Proposition 2) which, by Lemma 8, is $\bigcup_{A \in \mathscr{A}} A = E$ (by B 3).

5. Filters. In a topological space the sequence (x_n) is said to *converge* to a if to every neighbourhood U of a corresponds a positive integer k with $x_n \in U$ for all $n \geqslant k$; we then write $x_n \to a$. Many of the topological properties of metric spaces can conveniently be described in terms of sequences. For example, $a \in \bar{A}$ if and only if there is a sequence of points of A converging to a; the mapping f is continuous at a if and only if $f(x_n) \to f(a)$ whenever $x_n \to a$. These properties are in default for a general topological space, unless the notion of a sequence is also generalised. One way of doing this is to use filters.

Let E be any set. A non-empty set $\mathscr{F}$ of non-empty subsets of E is called a *filter* if it satisfies:

F 1: *if* $A \in \mathscr{F}$ *and* $B \in \mathscr{F}$, *then* $A \cap B \in \mathscr{F}$;

F 2: *if* $A \in \mathscr{F}$ *and* $A \subseteq B$, *then* $B \in \mathscr{F}$.

For example, if A is a fixed subset of E, the set of all subsets containing A is a filter; a more illuminating example of a filter is the set of all neighbourhoods of a fixed point of a topological space. Like the set of neighbourhoods of a point, a filter can be generated by a base. A non-empty set $\mathscr{B}$ of non-empty subsets of the set E is called a *filter base* if it satisfies:

FB: *if* $A \in \mathscr{B}$ *and* $B \in \mathscr{B}$, *there is some* $C \in \mathscr{B}$ *with* $C \subseteq A \cap B$.

The set $\mathscr{F}$ of all sets containing a set of $\mathscr{B}$ is then a filter, called the filter *generated* by $\mathscr{B}$.

If f maps E into F and $\mathscr{B}$ is a filter base (or a filter) in E then $f(\mathscr{B})$ is a filter base in F (for $f(A \cap B) \subseteq f(A) \cap f(B)$).

Suppose now that E is a topological space. The filter (or filter base) $\mathscr{F}$ is said to *converge* to a if every neighbourhood of a contains some set of $\mathscr{F}$. We then write $\mathscr{F} \to a$. When $\mathscr{F}$ is a filter (so that F 2 holds), $\mathscr{F} \to a$ if and only if every neighbourhood of a belongs to $\mathscr{F}$. Thus, if $\mathscr{F} \to a$, then $a \in \bar{A}$ for every $A \in \mathscr{F}$. For if U is a neighbourhood of a, $U \in \mathscr{F}$ and so $U \cap A \in \mathscr{F}$ by F 1; hence $U \cap A \neq \emptyset$, since the empty set was expressly excluded from every filter.

PROPOSITION 9. *In a separated topological space a filter cannot converge to more than one point.*

Proof. Suppose that the filter $\mathscr{F}$ converges to two distinct points a and b. There are disjoint neighbourhoods U of a and V of b, the space being separated. Then $U \in \mathscr{F}$ and $V \in \mathscr{F}$ and so $\emptyset = U \cap V \in \mathscr{F}$ (by F 1) which is not possible.

The condition for convergence can be expressed in another way. Given two filters $\mathscr{F}$ and $\mathscr{G}$, we say that $\mathscr{F}$ is *finer* than $\mathscr{G}$, or that $\mathscr{F}$ is a *refinement* of $\mathscr{G}$, if $\mathscr{G} \subseteq \mathscr{F}$. Then $\mathscr{F} \to a$ if and only if $\mathscr{F}$ is a refinement of the filter of neighbourhoods of a. If $\mathscr{F}$ converges to a, so also does any refinement of $\mathscr{F}$, and $\mathscr{F}$ remains convergent to a under any coarser topology.

We must justify the assertion that filters generalise sequences. The passage from a sequence (x_n) to a filter can be made by putting $X_n = \{x_i: i \geqslant n\}$; the set $\{X_n\}$ is then a filter base. The filter generated by it is called the *elementary filter* associated with the sequence, and we have the comforting result that a sequence converges to a if and only if the associated elementary filter converges to a. Refining a filter is analogous to selecting a subsequence. Let $\mathscr{F}$ and $\mathscr{G}$ be the elementary filters associated with the sequences (x_n) and (y_n); if (x_n) is a subsequence of (y_n) then $\mathscr{F}$ is finer than $\mathscr{G}$. Conversely, if $\mathscr{F}$ is finer than $\mathscr{G}$ then the sequence (x_n) comes close to being a subsequence of (y_n); precisely, there is some integer k with $x_n = y_{r(n)}$ for $n \geqslant k$ and $r(n) \to \infty$ as $n \to \infty$.

A filter which has no proper refinement (i.e. no refinement other than itself) is called an *ultrafilter*.

THEOREM 4. *Every filter has a refinement which is an ultrafilter.*

Proof. Apply the maximal axiom (Ch. I, § 1) to the set of all filters on E; if $\mathscr{F}$ is a filter on E, the (trivial) chain $\{\mathscr{F}\}$ is contained in a maximal chain. Let $\mathscr{G}$ be the union of the filters in this maximal chain. Then the filter axioms are easily verified for $\mathscr{G}$. Also $\mathscr{G}$ is a refinement of $\mathscr{F}$. Finally, $\mathscr{G}$ is clearly maximal, for any proper refinement of $\mathscr{G}$ could be added to the maximal chain and and thus contradict the maximality.

Ultrafilters have the following useful property:

PROPOSITION 10. *If $\mathscr{F}$ is an ultrafilter in E and A is a subset of E, then either $A \in \mathscr{F}$ or $\sim A \in \mathscr{F}$.*

Proof. If $A \notin \mathscr{F}$, then $\mathscr{F}$ contains no subset of A, and so $B \cap \sim A \neq \emptyset$ for each $B \in \mathscr{F}$. Then the sets $B \cap \sim A$, as B runs through $\mathscr{F}$, form the base of a filter finer than $\mathscr{F}$, and thus identical with $\mathscr{F}$. Hence $\sim A \in \mathscr{F}$.

COROLLARY. *If $\mathscr{F}$ is an ultrafilter and if $\bigcup_{1 \leqslant i \leqslant n} A_i \in \mathscr{F}$, then $A_i \in \mathscr{F}$ for some i.*

For if not, $\sim A_i \in \mathscr{F}$ for each i, and so

$$\emptyset = \Big(\bigcup_{1 \leqslant i \leqslant n} A_i\Big) \cap \Big(\sim \bigcup_{1 \leqslant i \leqslant n} A_i\Big) = \Big(\bigcup_{1 \leqslant i \leqslant n} A_i\Big) \cap \bigcap_{1 \leqslant i \leqslant n} (\sim A_i) \in \mathscr{F}.$$

The condition for a set to be compact can be expressed elegantly in terms of filter convergence:

PROPOSITION 11. *The following conditions on a subset A of a topological space are equivalent:*

(i) *A is compact;*

(ii) *if a family of closed non-empty sets has an intersection not meeting A, then so has some finite subfamily;*

(iii) *every filter to which A belongs has a refinement convergent to a point of A;*

(iv) *every ultrafilter to which A belongs is convergent to a point of A.*

Proof. The statements (i) and (ii) are shown to be equivalent by taking complements. Clearly (iii) implies (iv) and, by Theorem 4, (iv) implies (iii).

Suppose that (ii) holds, and that $\mathscr{F}$ is a filter with $A \in \mathscr{F}$. Then there is some $a \in A$ with

$$a \in \bigcap_{B \in \mathscr{F}} \bar{B}.$$

For otherwise, by (ii), there is a finite number of sets $B_1, B_2, \ldots, B_n$ with

$$A \cap \bigcap_{1 \leqslant i \leqslant n} \overline{B_i} = \emptyset,$$

which is impossible since $A \in \mathscr{F}$ and $\bar{B}_i \in \mathscr{F}$ for $1 \leqslant i \leqslant n$. Now every neighbourhood U of a meets every $B \in \mathscr{F}$ and so the sets $U \cap B$ form the base of a filter finer than both $\mathscr{F}$ and the neighbourhood filter at a, which is therefore a refinement of $\mathscr{F}$ convergent to a. Conversely, if (iii) holds, and if $\{B_\gamma\}$ is a family of closed sets with

$$A \cap \bigcap_{1 \leqslant i \leqslant n} B_{\gamma(i)} \neq \emptyset,$$

then these latter sets form the base of a filter to which A belongs. There is a refinement convergent to $a \in A$ and so $a \in \bar{B}_\gamma = B_\gamma$ for each γ. Hence

$$A \cap \bigcap_{\gamma} B_\gamma \neq \emptyset.$$

COROLLARY. *If $\mathscr{C}$ is a chain of closed non-empty subsets of a compact set A then* $\bigcap_{C \in \mathscr{C}} C \neq \emptyset$.

This is a consequence of (ii), since any finite intersection of sets of a chain is equal to the smallest of them.

6. Completeness. We now return to the study of convex spaces. In a convex space the filter $\mathscr{F}$ is called a *Cauchy filter* if, for each neighbourhood U, $\mathscr{F}$ contains a set small of order U.

Every convergent filter is a Cauchy filter, for if $\mathscr{F} \to a$ and U is an absolutely convex neighbourhood there is some $A \in \mathscr{F}$ with $A \subseteq a + \frac{1}{2}U$ and then A is small of order U. If, conversely, every Cauchy filter is convergent, the space E is called complete; more generally, the subset A of E is called *complete* if every Cauchy filter to which A belongs is convergent to a point of A. In view of Theorem 4, for A to be complete it is sufficient that the condition be fulfilled by Cauchy ultrafilters.

LEMMA 9. (i) *Any scalar multiple or any closed subset of a complete set is complete;*

(ii) *any finite union of complete sets is complete;*

(iii) *a complete subset of a separated space is closed.*

Proofs. (i) The first part is immediate. If A is complete and B a closed subset of A, then if B belongs to a Cauchy filter $\mathscr{F}$ so does A. Hence $\mathscr{F} \to a \in A$. But $B \in \mathscr{F}$ and so $a \in \bar{B} = B$. (ii) This part is a consequence of the Corollary to Proposition 10. (iii) If A is complete and $a \in \bar{A}$, the sets $(a+U) \cap A$ form the base of a filter $\mathscr{F}$ convergent to a. Then $\mathscr{F}$ is Cauchy and $A \in \mathscr{F}$ and so $\mathscr{F}$ converges to a point of A; this point must be a, the space being separated (Prop. 9). Thus $a \in A$; hence A is closed.

Completeness is exactly the property required to convert a precompact set into a compact one. In the first place, parallel to Proposition 11 we have:

LEMMA 10. *The following conditions on a subset A of a convex space are equivalent:*

(i) *A is precompact;*

(ii) *every filter to which A belongs has a Cauchy refinement;*

(iii) *every ultrafilter to which A belongs is Cauchy.*

Proof. The equivalence of (ii) and (iii) is a result of Theorem 4. Suppose that A is precompact. If $\mathscr{F}$ is an ultrafilter with $A \in \mathscr{F}$ and if U is any neighbourhood, then

$$A = \bigcup_{1 \leqslant i \leqslant n} A_i$$

with each A_i small of order U. By the Corollary of Proposition 10, there is some i with $A_i \in \mathscr{F}$ and so $\mathscr{F}$ is Cauchy. Suppose finally that (ii) holds but that A is not precompact. Then for some neighbourhood U, A has no finite covering by sets small of order U; hence, as B runs through the subsets of E that have a finite covering by sets small of order U, the sets $A \cap (\sim B)$ form the base of a filter $\mathscr{F}$ to which A belongs. This has a Cauchy refinement $\mathscr{G}$ and so there is some $C \in \mathscr{G}$ with C small of order U. But $A \cap (\sim C) \in \mathscr{F}$ and thus $\sim C \in \mathscr{F} \subseteq \mathscr{G}$; hence

$$\emptyset = C \cap (\sim C) \in \mathscr{G},$$

which is impossible.

Theorem 5. *A subset of a convex space is compact if and only if it is precompact and complete.*

Proof. The compactness of a precompact complete set follows from Proposition 11 and Lemma 10. Conversely, a compact set is certainly precompact, from the definitions; its completeness results from Proposition 11.

Corollary. *The closed (absolutely) convex envelope of a compact subset of a complete convex space is compact.*

For it is precompact (Th. 3, Cor. 2) and complete (Lemma 9 (i)).

The idea of a Cauchy filter is of course not unconnected with the familiar concept of a Cauchy sequence in a metric space (where (x_n) is Cauchy if $d(x_n, x_m) \to 0$ as $n, m \to \infty$). It can in fact be generalised to wider classes of topological spaces; in particular, a sequence (x_n) in a convex space is called a *Cauchy sequence* if for each neighbourhood U there is some integer k with $x_n - x_m \in U$ for all $n \geqslant k$ and $m \geqslant k$. With this definition, a sequence is Cauchy if and only if its associated elementary filter is Cauchy.

Proposition 12. *In a metrisable convex space, a set A is complete if and only if every Cauchy sequence of points of A is convergent to a point of A.*

Proof. Clearly completeness implies the convergence of every Cauchy sequence. Conversely, let $\mathscr{F}$ be a Cauchy filter with $A \in \mathscr{F}$. For each n, there is a set $A_n \in \mathscr{F}$ with diameter $d(A_n) < 1/n$. Let $a_n \in A_n \cap A$. Then (a_n) is a Cauchy sequence and so is convergent to a point $a \in A$. Any neighbourhood of a contains some A_n and thus $\mathscr{F} \to a$. Hence A is complete.

Complete metrisable convex spaces are called *Fréchet spaces*; complete normed spaces are called *Banach spaces*. (See Ch. VI, § 3.)

The scalar field of real or complex numbers is complete, by Cauchy's classical convergence principle. From Theorem 5 we can then deduce the well-known fact that a set of real or complex numbers is compact if and only if it is closed and bounded. For it is complete if and only if it is closed (Lemma 9 (i), (iii)) and precompactness is clearly equivalent to boundedness.

We can also deduce the completeness of another important convex space. First, however, we notice that the idea of filter convergence in a convex space enables us to justify the use of the phrase 'topology of uniform convergence' (see § 2). Suppose that (E, E') is a dual pair and that E' has the topology defined by a set $\mathscr{A}$ of bounded subsets of E. Let $\mathscr{F}'$ be a filter in E'. Then $\mathscr{F}' \to a' \in E'$ if and only if $\langle x, \mathscr{F}'\rangle \to \langle x, a'\rangle$ uniformly on each $A \in \mathscr{A}$. In particular, $\mathscr{F}' \to a'$ in $\sigma(E', E)$ if and only if

$$\langle x, \mathscr{F}'\rangle \to \langle x, a'\rangle$$

for each $x \in E$. We are now ready to prove:

Proposition 13. *Under the topology $\sigma(E^*, E)$, E^* is complete.*

Proof. Let $\mathscr{F}^*$ be a Cauchy filter in E^*. Then for each $x \in E$ and each $\epsilon > 0$, there is some $A^* \in \mathscr{F}^*$ for which $\langle x, A^*\rangle$ has diameter less than ϵ. Thus $\langle x, \mathscr{F}^*\rangle$ is the base of a Cauchy filter of scalars. The scalar field being complete, there is some $f(x)$ with

$$\langle x, \mathscr{F}^*\rangle \to f(x).$$

We show that f is a linear form on E. For each $x, y \in E$ and $\epsilon > 0$, there are sets $A^* \in \mathscr{F}^*$ with $|\langle x, x^*\rangle - f(x)| < \frac{1}{2}\epsilon$ for all $x^* \in A^*$, and $B^* \in \mathscr{F}^*$ with $|\langle y, x^*\rangle - f(y)| < \frac{1}{2}\epsilon$ for all $x^* \in B^*$. Then $C^* = A^* \cap B^* \in \mathscr{F}^*$ and $|\langle x+y, x^*\rangle - (f(x)+f(y))| < \epsilon$ for all $x^* \in C^*$. Hence $\langle x+y, \mathscr{F}^*\rangle \to f(x)+f(y)$, i.e. $f(x+y) = f(x)+f(y)$. Similarly $f(\lambda x) = \lambda f(x)$. Hence there is some $x^* \in E^*$ with $\langle x, \mathscr{F}^*\rangle \to \langle x, x^*\rangle$ for all $x \in E$. Thus $\mathscr{F}^* \to x^*$ in $\sigma(E^*, E)$.

7. The Mackey–Arens theorem.

The two theorems of this section are of fundamental importance in the theory of convex spaces.

Theorem 6. *If E is a separated convex space with dual E' and U is a neighbourhood of the origin then U^0 is $\sigma(E', E)$-compact.*

Proof. Give E^* the topology $\sigma(E^*, E)$. Since U is an absorbent set in E, U^0 is bounded and therefore precompact (Prop. 6). Also E^* is complete (Prop. 13) and U^0 closed (Ch. II, Prop. 9 (i)) and so U^0 is complete (Lemma 9 (i)). Hence U^0 is compact (Th. 5). But $U^0 \subseteq E'$ and the topologies $\sigma(E', E)$ and $\sigma(E^*, E)$ coincide on E', and so U^0 is $\sigma(E', E)$-compact (Lemma 5).

COROLLARY 1. *If, under the conditions of the theorem, A' is equicontinuous then $\overline{A'}$ is $\sigma(E', E)$-compact.*

COROLLARY 2. *The unit ball in the dual E' of a normed space E is $\sigma(E', E)$-compact.*

For it is the polar of the unit ball in E.

THEOREM 7. (Mackey–Arens theorem.) *Suppose that (E, E') is a dual pair, and that, under a topology ξ, E is a separated convex space. Then E has dual E' under ξ if and only if ξ is a topology of uniform convergence on a set of absolutely convex $\sigma(E', E)$-compact subsets of E'.*

Proof. If E has dual E' under ξ, then ξ is the topology of uniform convergence on the sets U^0, as U runs through the neighbourhoods under ξ, and each U^0 is absolutely convex and, by Theorem 6, $\sigma(E', E)$-compact.

Conversely, if ξ is such a topology, then Proposition 8 applies, with E and E' interchanged and with the topology $\sigma(E', E)$ on E', and shows that the dual of E is E'.

This theorem shows that there is a finest topology of the dual pair (E, E'), namely the topology of uniform convergence on the set of all absolutely convex $\sigma(E', E)$-compact subsets of E'. This topology is denoted by $\tau(E, E')$ and sometimes called the *Mackey topology*. It is clearly coarser than $\beta(E, E')$, the finest polar topology; if, under $\beta(E, E')$, the dual of E is E', then $\beta(E, E')$ is identical with $\tau(E, E')$.

In Chapter II, Proposition 13, we proved that a continuous linear mapping is also weakly continuous. The restricted converse, promised there, is:

PROPOSITION 14. *If E and F are separated convex spaces, and if E has dual E' and topology $\tau(E, E')$, then every weakly continuous linear mapping of E into F is also continuous.*

Proof. Let V be a closed absolutely convex neighbourhood in F. Then, by Theorem 6, V^0 is $\sigma(F', F)$-compact. Since the transpose t' of the weakly continuous linear mapping t is weakly continuous (Ch. II, Prop. 12, Corollary), $t'(V^0)$ is $\sigma(E', E)$-compact (Prop. 7). Hence its polar in E is a neighbourhood under $\tau(E, E')$. But, by Chapter II, Lemma 6, applied with t and t' interchanged, $(t'(V^0))^0 = t^{-1}(V^{00}) = t^{-1}(V)$, since V is closed and absolutely convex. Thus t is continuous.

SUPPLEMENT

(1) *Completeness of specific spaces.* Many of the spaces listed in the Supplement to Chapter I are complete.

If S is a separated compact or locally compact space, the space $\mathscr{C}(S)$, under the topology of compact convergence (Ch. I, Suppl. 2 *a*, *b*), is complete. (For a Cauchy sequence or filter converges to some function, uniformly on each compact set, and so uniformly on a neighbourhood of each point of the locally compact space S. Hence the limit function is continuous.) If S is a separated locally compact space, the union of a sequence of compact sets (but not itself compact), the space $\mathscr{K}(S)$ under the inductive limit topology is complete. (For $\mathscr{K}(S)$ is then the strict inductive limit of a sequence of complete spaces: see Ch. VII, Prop. 3.) On the other hand, $\mathscr{K}(S)$ is not complete under the topology of uniform convergence; its completion (see Ch. VI, § 1) is the space of continuous functions vanishing at infinity. (The continuous function x *vanishes at infinity* if for each $\epsilon > 0$ there is a compact set A with $|x(t)| < \epsilon$ for $t \in \sim A$.)

The spaces $\mathscr{D}$, $\mathscr{E}$ and $\mathscr{S}$ of indefinitely differentiable functions (Ch. I, Suppl. 3) are all complete; thus $\mathscr{E}$ and $\mathscr{S}$ are Fréchet spaces.

The completeness of the space $\mathscr{H}(D)$ of holomorphic functions (Ch. I, Suppl. 4) is a consequence of a theorem of Weierstrass or of Morera's converse of Cauchy's theorem. For a Cauchy sequence of holomorphic functions converges uniformly on each compact set to a function which is certainly continuous, and so, by either of these theorems, also holomorphic. Thus $\mathscr{H}(D)$ is also a Fréchet space.

The normed sequence and integration spaces (Ch. I, Suppls. 5

and 6) are all complete. (The completeness of $\mathscr{L}^2$ is the Riesz–Fischer theorem: see F. Riesz and B. Sz.-Nagy, *Functional analysis* (1956), II, §28.)

(2) *Finest convex topology.* The finest convex topology (Ch. I, Suppl. 8) on a vector space E is $\tau(E, E^*)$, since the dual is E^* (Ch. II, Suppl. 7). In E, only finite-dimensional sets can be bounded. (For if (x_n) is an infinite sequence of linearly independent elements, extend $\{x_n\}$ to form a base B and define x^* by $\langle x_n, x^*\rangle = n$, $\langle x, x^*\rangle = 0$ for all other $x \in B$, and by linearity elsewhere. Then x^* is unbounded on $\{x_n\}$ and so $\{x_n\}$ is not a bounded set.) Hence all the polar topologies on E^* coincide. In particular, $\beta(E^*, E) = \tau(E^*, E) = \sigma(E^*, E)$. Also E can only be normable under $\tau(E, E^*)$ if it is finite-dimensional. The space E is complete under $\tau(E, E^*)$ (see Ch. V, Suppl. 1 and Prop. 23: this special case can be proved directly and the details are simpler, but the method is essentially the same).

(3) *Tychonoff's theorem.* There is a theorem, due to Tychonoff, about product spaces which can be used to give an alternative proof of Theorem 6. It states that a product of compact sets is compact (see N. Bourbaki, *Éléments de mathématique, Topologie générale* (1940), Ch. I, §10, Th. 2). Now if (E, E') is a dual pair, then under the topology $\sigma(E', E)$, E' can be regarded as a subspace of the product Φ^E (see Ch. V, Suppl. 1). For each absolutely convex neighbourhood U in E, with gauge p, the set of $\phi \in \Phi^E$ with $|\phi(x)| \leqslant p(x)$ for all x is a product of intervals (discs) (one for each x), and so, by Tychonoff's theorem, is compact. Now ϕ is linear if and only if for each $x, y \in E$ and $\lambda, \mu \in \Phi$,

$$t(\phi) = \phi(\lambda x + \mu y) - \lambda\phi(x) - \mu\phi(y) = 0.$$

Since each such function t is continuous, the subset of Φ^E consisting of linear forms on E is the intersection of the closed sets $t^{-1}(0)$ and is therefore closed. Hence the set U^0 is a closed subset of a compact set and so is compact (under the product topology on Φ^E and so also under $\sigma(E', E)$).

CHAPTER IV

BARRELLED SPACES AND THE BANACH–STEINHAUS THEOREM

One of the most powerful theorems of functional analysis, the Banach–Steinhaus theorem, asserts that a set of continuous linear mappings that is bounded at each point of a Banach space is bounded uniformly on the unit ball. So valuable is this principle that the convex spaces for which its natural extension is still valid merit special study. Here called barrelled spaces, they are introduced and characterised in the first section of this chapter, and the next section contains the principle of uniform boundedness that holds for them (Theorem 3). This, together with Theorem 2, which asserts that every Fréchet space is barrelled, reconstitute the Banach–Steinhaus theorem. The notion of a barrelled space is helpful in many other contexts, as we shall see in later chapters; it occurs naturally in the discussion of reflexivity contained in the last section of this chapter.

1. Barrelled spaces. In a convex space, a subset is called a *barrel* if it is absolutely convex, absorbent and closed. Every convex space has a neighbourhood base consisting of barrels; a convex space is called *barrelled* if every barrel is a neighbourhood.

If (E, E') is a dual pair, the closure of an absolutely convex set in E is the same for all topologies of (E, E'), by Proposition 8 of Chapter II, and therefore the property of being a barrel in E depends only on the dual pair (E, E').

PROPOSITION 1. *Let E be a separated convex space with dual E'. Then the subset B of E is a barrel if and only if B is the polar of a $\sigma(E', E)$-bounded subset of E'.*

Proof. The polar of a $\sigma(E', E)$-bounded set is absolutely convex, closed (Ch. II, Props. 9 and 8), and absorbent (Ch. III, Lemma 2). Conversely, if B is a barrel then $B = B^{00}$ (Ch. II, Th. 4, Cor. 1) and B^0 is $\sigma(E', E)$-bounded (Ch. III, Lemma 2).

Corollary 1. *If E is a separated convex space with dual E' then E is barrelled if and only if every $\sigma(E', E)$-bounded subset of E' is equicontinuous, i.e. if and only if the topology of E is $\beta(E, E')$.*

Corollary 2. *If E is a separated barrelled space with dual E', then E has the topology $\tau(E, E')$.*

For then $\beta(E, E')$ is a topology of the dual pair (E, E').

Corollary 3. *If E is a separated barrelled space with dual E', the closed [absolutely] convex envelope of every $\sigma(E', E)$-compact subset of E' is $\sigma(E', E)$-compact.*

For the absolutely convex envelope of a $\sigma(E', E)$-compact set is $\sigma(E', E)$-bounded and so equicontinuous (Cor. 1); its closure, and any closed subset, are therefore $\sigma(E', E)$-compact.

Lemma 1. *In a convex space, a barrel absorbs every convex compact set.*

Proof. Let B be a barrel and A a convex compact set. It is sufficient to show the existence of a positive integer n, a neighbourhood U and a point x in A with

$$A \cap (x + U) \subseteq nB, \quad \text{i.e.} \quad (A - x) \cap U \subseteq nB - x.$$

For $A - x$ is bounded and so $A - x \subseteq \lambda U$ for some $\lambda \geqslant 1$; also $o \in A - x$ and therefore $A - x \subseteq \lambda(A - x)$. Hence

$$A - x \subseteq \lambda(A - x) \cap \lambda U \subseteq \lambda(nB - x),$$

and so
$$A \subseteq \lambda nB - (\lambda - 1)x \subseteq \mu B$$

for some μ since B is absorbent.

Suppose then that no n, U, x can be found to satisfy the condition $A \cap (x + U) \subseteq nB$. Then, taking $n = 1$, any $x_0 \in A$ and any open neighbourhood U_0, there is some

$$x_1 \in A \cap (x_0 + U_0) \cap ({\sim} B).$$

Now $(x_0 + U_0) \cap ({\sim} B)$ is open and so there is some open U_1 with $x_1 + \overline{U_1} \subseteq (x_0 + U_0) \cap ({\sim} B)$. Take $n = 2$, $x = x_1$, $U = U_1$; there is some $x_2 \in A \cap (x_1 + U_1) \cap ({\sim} 2B)$. Now $(x_1 + U_1) \cap ({\sim} 2B)$ is open and so there is some open U_2 with $x_2 + \overline{U_2} \subseteq (x_1 + U_1) \cap ({\sim} 2B)$. Take $n = 3$, $x = x_2$, $U = U_2$ and so on. Then $(A \cap (x_n + \overline{U_n}))$ is a decreasing sequence of closed non-empty sets. Since A is compact, they have a common point $a \in A$ (Ch. III, Prop. 11). For each n, $a \notin nB$, and so B is not absorbent, which is a contradiction.

COROLLARY. *In a separated convex space, a barrel absorbs every complete convex bounded set.*

Proof. Let A be a complete convex bounded set and B a barrel. If $a \in A$, and B does not absorb $A_0 = A - a$, there is a sequence (x_n) of points of A_0 with $x_n \notin n^2 B$, and so the sequence $(n^{-1}x_n)$ is not absorbed by B. But $n^{-1}x_n \to o$, for, if U is a neighbourhood, there is some $m > 0$ with $A_0 \subseteq mU$ and then $n^{-1}x_n \in U$ for $n \geqslant m$. Hence the set consisting of the points $n^{-1}x_n$ together with o is (pre)compact and so its closed convex envelope C is precompact (Ch. III, Th. 3, Cor. 2). Since C is a closed subset of A_0, it is complete and therefore compact. But C is not absorbed by B, which contradicts the lemma. Thus B absorbs A_0 and so A.

THEOREM 1. *The same sets are bounded in every topology of a dual pair.*

Proof. If ξ is any topology of the dual pair (E, E') the ξ-bounded sets are certainly $\sigma(E, E')$-bounded. Conversely, let A be a $\sigma(E, E')$-bounded set and U a closed absolutely convex ξ-neighbourhood. By Proposition 1, A^0 is a barrel in E' (under $\sigma(E', E)$) and U^0 is absolutely convex and $\sigma(E', E)$-compact (Ch. III, Th. 6); hence A^0 absorbs U^0 (Lemma 1). Therefore (Ch. II, Prop. 9) U^{00} absorbs A^{00}. But $U^{00} = U$ (Ch. II, Th. 4, Cor. 1) and $A \subseteq A^{00}$; hence U absorbs A. Thus A is ξ-bounded.

COROLLARY. *If E is a metrisable convex space with dual E', then E has the topology $\tau(E, E')$.*

Proof. Let (U_n) be a base of neighbourhoods, with $U_{n+1} \subseteq U_n$ for all n, in the metrisable topology ξ, and V any $\tau(E, E')$-neighbourhood. If V is not a ξ-neighbourhood there are points $x_n \in U_n$ with $x_n \notin nV$. Then $\{x_n\}$ is ξ-bounded (in fact $x_n \to o$ in ξ) and so by the theorem $\{x_n\}$ is $\tau(E, E')$-bounded. Hence there is some $\mu > 0$ with $x_n \in \lambda V$ for $|\lambda| \geqslant \mu$ and all n, which is a contradiction.

Another consequence of this theorem is that the norm topology on the dual E' of a normed space E is $\beta(E', E)$ (cf. Ch. III, § 2).

THEOREM 2. *Every Fréchet space (i.e. complete metrisable convex space) is barrelled. In particular, every Banach space is barrelled.*

Proof. Let (U_n) be a base of neighbourhoods, with $U_{n+1} \subseteq U_n$ for all n, and B a barrel in E. If B is not a neighbourhood there is a sequence (x_n) with $x_n \in U_n$ but $x_n \notin nB$. Then $x_n \to o$ and so $A = \{x_n\} \cup \{o\}$ is compact. Hence the closed convex envelope of A is compact (Ch. III, Th. 5, Corollary). Hence by Lemma 1 it is absorbed by B; thus there is some $\lambda > 0$ with $x_n \in \lambda B$ for all n, which is a contradiction.

We shall be able to enlarge the class of barrelled spaces later (Ch. v, Props. 6, 10 and 27).

2. Topologies on spaces of linear mappings. Let E and F be two convex spaces over the same (real or complex) field Φ and let L be the vector space of all continuous linear mappings of E into F. An extension of the method used in Chapter III to topologise E' (the special case of L when $F = \Phi$) can be applied to L. Let $\mathscr{A}$ be any set of bounded subsets of E and $\mathscr{V}$ a base of absolutely convex neighbourhoods in F. For each $A \in \mathscr{A}$ and each $V \in \mathscr{V}$ put $W_{A,V} = \{t : t(A) \subseteq V\}$. Then $W_{A,V}$ is absolutely convex, and it is also absorbent, because if $t \in L$, $t(A)$ is bounded (Ch. III, Prop. 1) and so there is some $\lambda > 0$ with $t(A) \subseteq \lambda V$ and then $t \in \lambda W_{A,V}$. Hence the sets $W_{A,V} (A \in \mathscr{A}, V \in \mathscr{V})$ define a convex topology on L. If, as is usual, the conditions B 1, B 2 and B 3 (Ch. III, § 2) are satisfied by $\mathscr{A}$, the sets $W_{A,V}$ form a base of neighbourhoods, and the topology is separated whenever F is. The topology is then called the *topology of $\mathscr{A}$-convergence*, or of *uniform convergence* on the sets of $\mathscr{A}$.

For each $A \in \mathscr{A}$ and each continuous seminorm q on F, put $q_A(t) = \sup\{q(t(x)) : x \in A\}$. Then q_A is a continuous seminorm on L; these seminorms determine the topology of $\mathscr{A}$-convergence.

The subset T of L is bounded in the topology of $\mathscr{A}$-convergence if and only if $\bigcup_{t \in T} t(A)$ is bounded in F for each $A \in \mathscr{A}$ (for $T \subseteq \lambda W_{A,V}$ if and only if $\bigcup_{t \in T} t(A) \subseteq \lambda V$).

The coarsest topology of $\mathscr{A}$-convergence arises when $\mathscr{A}$ is the set of finite subsets of E, and is called the topology of pointwise convergence. The subset T is pointwise bounded if and only if $T(x) = \{t(x) : t \in T\}$ is bounded for each $x \in E$. The finest is obtained by taking $\mathscr{A}$ to be the set of all bounded subsets of E.

If E and F are both normed spaces this is also a norm topology with

$$\|t\| = \sup\{\|t(x)\| : \|x\| \leqslant 1\}.$$

If T is an equicontinuous subset of L, then T is bounded in every topology of $\mathscr{A}$-convergence. For if V is a neighbourhood in F, there is some neighbourhood U in E with $t(U) \subseteq V$ for all $t \in T$; if $A \in \mathscr{A}$ there is some $\alpha > 0$ with $A \subseteq \lambda U$ for all $|\lambda| \geqslant \alpha$. Then $t(A) \subseteq \lambda V$ for all $|\lambda| \geqslant \alpha$. Conversely:

THEOREM 3. *Let E be a barrelled space and F a convex space. Then any pointwise bounded set of continuous linear mappings of E into F is equicontinuous.*

Proof. Let T be a pointwise bounded set of continuous linear mappings. If V is a closed absolutely convex neighbourhood in F, put $B = \bigcap_{t \in T} t^{-1}(V)$. Then B is absolutely convex and closed. It is also absorbent, for if $x \in E$, $T(x)$ is bounded and so there is some λ with $T(x) \subseteq \lambda V$; thus $x \in \lambda B$. Hence B is a barrel in the barrelled space E and is therefore a neighbourhood. But $t(B) \subseteq V$ for all $t \in T$ and so T is equicontinuous.

Taking F to be the scalar field, Corollary 1 of Proposition 1 shows that the theorem can hold only when E is barrelled.

COROLLARY 1. *Let E be a barrelled space and F a separated convex space. If (t_n) is a sequence of continuous linear mappings of E into F which is pointwise convergent to t_0, then t_0 is a continuous linear mapping and the convergence is uniform on every precompact subset of E.*

Proof. For each $x \in E$, $\{t_n(x)\}$ is bounded and so $\{t_n\}$ is equicontinuous. Hence if V is a closed absolutely convex neighbourhood in F, there is a neighbourhood U in E with $t_n(U) \subseteq V$ for all n. Then if $x \in U$,

$$t_0(x) = \lim_{n \to \infty} t_n(x) \in \overline{V} = V.$$

Thus t_0 is continuous (and clearly linear).

Let A be a precompact set. There are points $x_1, x_2, \ldots, x_k$ with

$$A \subseteq \bigcup_{1 \leqslant i \leqslant k} (x_i + \tfrac{1}{3}U)$$

and there are integers n_i with $t_n(x_i) - t_0(x_i) \in \frac{1}{3}V$ for $n \geqslant n_i$. Let $n_0 = \max_{1 \leqslant i \leqslant k} n_i$. Then for each $x \in A$, there is some i with $x - x_i \in \frac{1}{3}U$, and if $n \geqslant n_0$,

$$\begin{aligned} t_n(x) - t_0(x) &= t_n(x - x_i) + (t_n(x_i) - t_0(x_i)) - t_0(x - x_i) \\ &\in t_n(\tfrac{1}{3}U) + \tfrac{1}{3}V + t_0(\tfrac{1}{3}U) \subseteq V; \end{aligned}$$

thus the convergence on A is uniform.

COROLLARY 2. *Let E be a Banach space and F any normed space, and let T be a set of continuous linear mappings of E into F. If, for each $x \in E$,*

$$\sup_{t \in T} \|t(x)\| < \infty,$$

then

$$\sup_{t \in T} \|t\| < \infty.$$

If, for each $x \in E$, $t_n(x) \to t_0(x)$, then t_0 is a continuous linear mapping.

For a Banach space is barrelled (Th. 2).

3. The bidual and reflexivity. When we studied polar topologies in Chapter III, we showed that for each dual pair (E, E') there is a finest polar topology on E', namely the strong topology $\beta(E', E)$. This is the topology of uniform convergence on the bounded subsets of E. (After Theorem 1, there is no ambiguity in not specifying the topology on E in which the sets are to be bounded.) The dual E'' of E' under this topology is called the *bidual* of E; it is the union of the $\sigma(E'^*, E')$-closures of the bounded subsets of E (Ch. III, Prop. 2). The dual of E' under any polar topology lies between E and E''.

Since (E', E'') is a dual pair, the space E'' can be given various polar topologies, of which the finest is $\beta(E'', E')$, the topology of uniform convergence on the set of all $\sigma(E', E'')$-bounded (or equivalently $\beta(E', E)$-bounded) subsets of E'. We shall describe these sets as the *strongly bounded* subsets of E', to distinguish them from the $\sigma(E', E)$-bounded sets, which we shall call the *weakly bounded* subsets of E'. Since $\sigma(E', E'')$ is finer than $\sigma(E', E)$, every strongly bounded set is weakly bounded, but in general these two classes of sets are distinct. The strongly bounded subsets of E' are simply identified:

LEMMA 2. *If (E, E') is a dual pair, the subset A' of E' is strongly bounded if and only if its polar A'^0 in E absorbs bounded sets.*

Proof. The set A'^0 absorbs the bounded sets of E if and only if A' is absorbed by their polars, the $\beta(E', E)$-neighbourhoods.

COROLLARY. *If ξ is any topology of the dual pair (E, E'), every ξ-equicontinuous subset of E' is strongly bounded; every absolutely convex $\sigma(E', E)$-compact set is strongly bounded.*

For if A' is ξ-equicontinuous, its polar in E is a ξ-neighbourhood and so absorbs bounded sets. Also every absolutely convex $\sigma(E', E)$-compact set is $\tau(E, E')$-equicontinuous (Ch. III, § 7).

It follows from the Corollary that the set of all ξ-equicontinuous subsets of E' satisfies the conditions for defining a polar topology on E''. If $\mathscr{U}$ is a base of closed absolutely convex neighbourhoods for ξ, the bipolars U^{00}, taken in E'', of the sets of $\mathscr{U}$ form a base of neighbourhoods for this topology, which for this reason we denote by ξ^{00}. Since $U^{00} \cap E = U$, ξ^{00} induces ξ on E. Thus ξ^{00} can be thought of as being the result of 'lifting' the topology ξ from E to E''. In this way, if we are given a separated convex space E with topology ξ, dual E' and bidual E'', there are two natural topologies to be considered on E'', the topology ξ^{00} of equicontinuous convergence and the strong topology $\beta(E'', E')$. Of these, the second is always the finer. The identity mapping of E into E'' is always an isomorphism (into) when E'' has the topology ξ^{00}.

PROPOSITION 2. *Let E be a separated convex space with dual E' and with bidual E'' under the topology $\beta(E'', E')$. Then the identity mapping of E into E'' is an isomorphism if and only if every strongly bounded subset of E' is equicontinuous.*

Proof. Both conditions are equivalent to the identity of the topologies ξ^{00} and $\beta(E'', E')$, where ξ is the topology of E.

COROLLARY. *The identity mapping of E into E'' is an isomorphism (into) if E is barrelled.*

For this implies that every weakly bounded subset of E' is equicontinuous.

In Chapter V, Proposition 9, Corollary, another condition is given for the identity mapping of E into E'' to be an isomorphism, one which is satisfied whenever E is metrisable.

Though in general weakly and strongly bounded subsets of E' are distinct, there are certain conditions under which they are identical. One such is that E be barrelled. Another is given in:

PROPOSITION 3. *Let E be a complete separated convex space with dual E'. Then every weakly bounded subset of E' is strongly bounded.*

Proof. If A' is a weakly bounded subset of E', its polar A'^0 in E is a barrel (Prop. 1) and so absorbs complete convex bounded sets (Lemma 1, Corollary). But, since E is complete, every bounded subset of E is contained in a complete convex bounded set (its closed convex envelope). Hence A'^0 absorbs all bounded sets and so, by Lemma 2, A' is strongly bounded.

Suppose that (E, E') is a dual pair and that the bidual of E is just E itself. We shall then call the dual pair (E, E') *reflexive*.

PROPOSITION 4. *The dual pair (E, E') is reflexive if and only if every bounded set in E is contained in a weakly compact set.*

Proof. Clearly (E, E') is reflexive if and only if $\beta(E', E)$ is a topology of the dual pair (E', E) and so (Ch. III, Th. 7) if and only if every bounded set in E is a subset of an absolutely convex weakly compact set.

When E is a separated convex space with dual E', it is natural to call E reflexive if it is isomorphic, in some sense, with its bidual. If we demand only that this isomorphism be an algebraic one, we obtain a condition which is equivalent to the reflexivity of the dual pair (E, E') as defined already. (The space E is then called semi-reflexive by some authors.) Instead we may make the more stringent demand that E be isomorphic with E'' topologically as well as algebraically; we reserve the word *reflexive* to describe this latter case. Putting together Propositions 2 and 4, we obtain a characterisation of reflexive spaces:

PROPOSITION 5. *Let E be a separated convex space with dual E'. Then E is reflexive if and only if every bounded set in E is contained in a weakly compact set and every bounded set in E' is equicontinuous.*

(The first part of this condition for reflexivity ensures that the weakly and strongly bounded subsets of E' coincide. Hence the second part may refer either to weakly or to strongly bounded sets.)

Corollary 1. *A separated convex space is reflexive if and only if it is barrelled and every bounded set is contained in a weakly compact set.*

Corollary 2. *If a separated convex space is reflexive, so is its dual under the strong topology.*

Corollary 3. *A normed space is reflexive if and only if its unit ball is weakly compact.*

For in the dual of a normed space every strongly bounded set is absorbed by the unit ball and so is equicontinuous.

SUPPLEMENT

(1) *Category.* Anyone who has met this topological notion will recognise that the proof of Lemma 1 is essentially a category-type argument. A subset of a topological space E is called *nowhere dense* (*rare*) if its closure has no interior points. The subset A of E is said to be *of the first category* (*meagre*) *in* E if it can be expressed as the union of a sequence of nowhere dense subsets; otherwise it is *of the second category in* E. Thus A is of the second category in E if, whenever A is contained in the union of a sequence of closed subsets of E, at least one of them has an interior point. A *Baire space* E is one in which every (non-empty) open set is of the second category in E, and Baire's theorem states that every complete metric space, or every locally compact regular space, is a Baire space. If we assume this theorem, the proof of Lemma 1 is easy:

$$A = \bigcup_{n=1}^{\infty} A \cap nB,$$

and so at least one of the closed sets $A \cap nB$ contains an interior point (of the compact regular space A). There is therefore an integer n, a neighbourhood U in E and a point x in A with $A \cap (x+U) \subseteq nB$, and the proof continues as before.

A convex space is a Baire space if (and only if) it is of the second category (in itself). (For if the open set A were of the first category in E, then, since

$$E = \bigcup_{n=1}^{\infty} n(A-x)$$

for any $x \in A$, E would also be of the first category.) A vector subspace of a convex space E is either nowhere dense or dense in E; a vector subspace of the second category in E is therefore dense in E. It is also barrelled (in the topology induced on it as a subspace of E). (For if B is a barrel in the vector subspace M of the second category in E, then

$$M \subseteq \bigcup_{n=1}^{\infty} n\bar{B},$$

and so one of the sets $n\bar{B}$ has an interior point. If $x+U \subseteq n\bar{B}$, where U is an absolutely convex neighbourhood, then

$$U = \tfrac{1}{2}(x+U) - \tfrac{1}{2}(x+U) \subseteq \tfrac{1}{2}n\bar{B} - \tfrac{1}{2}n\bar{B} = n\bar{B},$$

and so $\bar{B}$ is a neighbourhood. Hence $B = M \cap \bar{B}$ is a neighbourhood in M.) Thus, in particular, a convex Baire space is barrelled. On the other hand, there are barrelled spaces which are not Baire spaces. For if E is a vector space with a countable base, then under the topology $\tau(E, E^*)$, E is barrelled (see (3) below) but it is not a Baire space since it is the union of the sequence of closed subspaces spanned by the first n base elements. (This is a special case of a strict inductive limit of a sequence of closed subspaces (see Ch. VII, § 1) which is never a Baire space, but is barrelled if each subspace is (Ch. V, Prop. 6).) (See also Ch. VI, Suppl. 2.)

(2) *Montel spaces.* A separated barrelled space with the further property that its closed bounded subsets are compact is called a *Montel space*. The name derives from the fact that it is a theorem of Montel that shows that the space $\mathscr{H}(D)$ of holomorphic functions has this property. In the first place, $\mathscr{H}(D)$ is a Fréchet space and so barrelled; since it is a metric space, it is enough to prove that every bounded sequence contains a convergent subsequence. But this, interpreted in the topology on $\mathscr{H}(D)$ of uniform convergence on compact subsets of the domain D, is precisely the content of Montel's theorem.

A Montel space is reflexive (Prop. 5); moreover, the dual E' of a Montel space E is also a Montel space under the strong topology. For it is certainly barrelled, every bounded subset of its dual E being contained in a compact set, and for the same reason the strong topology coincides with the topology of compact convergence. But every bounded set in E' is equicontinuous, and equicontinuous sets are contained in sets compact under the topology of compact convergence (Ch. VI, Prop. 4). A normed space can only be a Montel space if its closed unit ball is compact and so only if it is finite dimensional (Ch. III, Th. 2).

Other examples of Montel spaces are the spaces $\mathscr{D}$, $\mathscr{E}$ and $\mathscr{S}$ of indefinitely differentiable functions on $]-\infty, \infty[$ (Ch. I, Suppl. 3), and so also their duals, the spaces $\mathscr{D}'$, $\mathscr{E}'$ and $\mathscr{S}'$ of distributions (Ch. II, Suppl. 3), under their appropriate strong topologies, and the dual of $\mathscr{H}(D)$ (Ch. II, Suppl. 4). (See also Ch. VII, Suppl. 1.)

(3) *Finest convex topology.* Under its finest convex topology $\tau(E, E^*)$, any vector space E is barrelled and, further, a Montel space. For a barrel, being absolutely convex and absorbent, is a neighbourhood, and since bounded sets are finite dimensional (Ch. III, Suppl. 2), any closed bounded set is compact. Hence E^* is barrelled under its strong topology $\beta(E^*, E) = \sigma(E^*, E)$, and also a Montel space. For, since E is barrelled, the closed bounded subsets of E^* are (weakly) compact.

CHAPTER V

INDUCTIVE AND PROJECTIVE LIMITS

Given a vector space and certain associated linear mappings, the question often arises of topologising the space in such a way as to make the mappings continuous. There are two cases according as the linear mappings are from other convex spaces into the given space or from the given space into other convex spaces. This chapter contains an account of inductive (direct) and projective (inverse) limit topologies, which provide the natural answers to these questions. Separate sections are devoted to the important special cases of quotients, products and direct sums of convex spaces. There is no central theorem in this chapter; the most important results express the permanence of certain desirable properties of convex spaces under formation of inductive or projective limits. For example, every inductive limit of barrelled spaces is barrelled and every product or direct sum of complete spaces is complete.

There is an imperfect duality between the notions of inductive and projective limits, well illustrated by the fact that every convex space is a projective limit of normed (or even Banach) spaces, whereas only certain convex spaces can be expressed as inductive limits of normed spaces. Those that can have various additional properties, such as completeness of their strong duals; they are studied in the third section of this chapter. A special case of inductive limits, having many properties not shared by the general inductive limits considered here, forms the subject matter of the first section of Chapter VII.

1. Quotient spaces. Let E be a vector space over Φ and let M be a vector subspace of E. Then the relation $x-y \in M$ is an equivalence on E, and the set of equivalence classes $X, Y, \ldots$ can be made into a vector space over Φ, called the *quotient of E by M*, and denoted by E/M. (If $X, Y \in E/M$ then $X+Y \in E/M$ and $\lambda X \in E/M$ for $\lambda \neq 0$; the definition of the algebraic operations is completed by putting $0.X = M$, the origin in E/M.) Given any element x of E, the equivalence class $k(x)$ to which x belongs is $x+M$, and k is a linear mapping, called the *canonical mapping* of E onto E/M.

If F is another vector space over Φ, any linear mapping t of E into F vanishing on M has a decomposition $t = u \circ k$, where u is a linear mapping of E/M into F; $u(X)$ is the common value of $t(x)$ for $x \in X$. In fact the mapping $t \to u$ is easily verified to be an isomorphism of the vector space of all linear mappings of E into F that vanish on M onto the vector space of all linear mappings of E/M into F.

If E is a convex space and $\mathscr{U}$ is a base of absolutely convex neighbourhoods, the sets $k(U)$ $(U \in \mathscr{U})$ form a base of neighbourhoods in a topology on E/M, called the *quotient topology*; under it, E/M is a convex space. Since $U \subseteq k^{-1}(k(U))$ for each $U \in \mathscr{U}$, k is continuous; in fact the quotient topology is clearly the finest topology on E/M under which k is continuous.

If p is the gauge of the absolutely convex neighbourhood U, the gauge q of $k(U)$ is given by

$$q(X) = \inf\{\lambda \colon \lambda \geqslant 0,\ X \in \lambda k(U)\}.$$

Now $X \in \lambda k(U)$ if and only if there is some $x \in X$ with $x \in \lambda U$, and so

$$q(X) = \inf_{x \in X} \inf\{\lambda \colon \lambda \geqslant 0,\ x \in \lambda U\}.$$

Hence

$$q(X) = \inf\{p(x) \colon x \in X\}.$$

Proposition 1. *Under the quotient topology, E/M is separated if and only if M is a closed vector subspace of E.*

Proof. If E/M is separated, the set consisting of the origin of E/M alone is closed and so its inverse image M by the continuous mapping k is a closed subset of E.

Next let M be closed in E, and let X be any point of E/M other than the origin. Then, for each $x \in X$, $x \notin M$ and so there is an absolutely convex neighbourhood U with $(x+U) \cap M = \emptyset$. Then $x \notin M+U$ and so $X \notin k(U)$. Thus E/M is separated.

If E is metrisable and M closed, then E/M is separated, by Proposition 1. Since E/M clearly has a countable base of neighbourhoods, it is metrisable.

If E is normed and M closed, E/M is also normed and

$$\|X\| = \inf\{\|x\| : x \in X\}.$$

If t is a linear mapping of the convex space E into the convex space F vanishing on the vector subspace M of E, we have seen that we can write $t = u \circ k$, where u maps E/M into F and k is the canonical mapping of E onto E/M. Now t is continuous if and only if, for each neighbourhood V in F,

$$t^{-1}(V) = (u \circ k)^{-1}(V) = k^{-1}(u^{-1}(V))$$

is a neighbourhood in E, i.e. $u^{-1}(V)$ is a neighbourhood in E/M. Thus t is continuous if and only if u is. A useful special case of these results is:

PROPOSITION 2. *Any linear mapping t of the convex space E into the convex space F has a decomposition $t = u \circ k$, where u is a* (1, 1) *linear mapping of $E/t^{-1}(o)$ into F and k is the canonical mapping of E onto $E/t^{-1}(o)$; t is continuous if and only if u is.*

It is easy to identify the dual of a quotient:

PROPOSITION 3. *If M is a vector subspace of the convex space E with dual E', then the dual of E/M is the polar M^0 of M in E'.*

Proof. The space of continuous linear forms on E/M is isomorphic to the space of continuous linear forms on E vanishing on M, i.e. to M^0.

We note, for use later, that the transpose k' of the canonical mapping k of E into E/M is the identity mapping of M^0 into E'.

In general, the quotient of a complete space E by a closed vector subspace M need not be complete, but it is for a Fréchet space E, as we shall prove in Chapter VI, Proposition 13, Corollary.

2. Inductive limits. Suppose that $(E_\gamma : \gamma \in \Gamma)$ is a family of convex spaces, all vector subspaces of a vector space E (as yet not topologised), with the property that their union spans E. It is natural to ask if their topologies can be pieced together to induce a topology on E. More specifically, can E be given a convex topology in such a way that a linear mapping defined on E will be continuous if and only if it is continuous on each E_γ? We prove the existence of such a topology below and study its properties. Instead of requiring that the E_γ be vector subspaces of E, it is sufficient to be given linear mappings u_γ of E_γ into E, by which the topology on E_γ can be transferred to $u_\gamma(E_\gamma)$; we then assume that $\bigcup_{\gamma \in \Gamma} u_\gamma(E_\gamma)$ spans E.

PROPOSITION 4. *For each $\gamma \in \Gamma$, let E_γ be a convex space and u_γ a linear mapping of E_γ into a vector space E, so that $\bigcup_{\gamma \in \Gamma} u_\gamma(E_\gamma)$ spans E. Then there is a finest convex topology on E under which all the u_γ are continuous. A base of neighbourhoods for this topology is formed by the set $\mathscr{U}$ of all absolutely convex subsets U of E, such that, for each γ, $u_\gamma^{-1}(U)$ is a neighbourhood in E_γ.*

Proof. If U is an absolutely convex neighbourhood in any topology on E making all the u_γ continuous, then each $u_\gamma^{-1}(U)$ is a neighbourhood in E_γ and so $U \in \mathscr{U}$. Now if $U \in \mathscr{U}$, $u_\gamma^{-1}(U)$ is absorbent in E_γ and so U absorbs all the points of $u_\gamma(E_\gamma)$; since $\bigcup_{\gamma \in \Gamma} u_\gamma(E_\gamma)$ spans E, U is absorbent in E. Hence $\mathscr{U}$ satisfies the conditions of Chapter I, Theorem 2, for a base of neighbourhoods for a convex topology on E and this is therefore the finest convex topology making each u_γ continuous.

COROLLARY. *If, for each $\gamma \in \Gamma$, $\mathscr{V}_\gamma$ is a base of absolutely convex neighbourhoods in E_γ, then the set $\mathscr{V}$ of absolutely convex envelopes of sets of the form $\bigcup_{\gamma \in \Gamma} u_\gamma(V_\gamma)$ with $V_\gamma \in \mathscr{V}_\gamma$ forms a base of neighbourhoods for E.*

Proof. By the proposition, the sets of $\mathscr{V}$ are neighbourhoods in E. Also, if U is any absolutely convex neighbourhood in E, $u_\gamma^{-1}(U)$ contains a neighbourhood $V_\gamma \in \mathscr{V}_\gamma$ and then the absolutely convex envelope of $\bigcup_{\gamma \in \Gamma} u_\gamma(V_\gamma)$ is a set of $\mathscr{V}$ contained in U. Thus $\mathscr{V}$ is a neighbourhood base for E.

The convex space E with this topology is called the *inductive limit* of the convex spaces E_γ by the mappings u_γ.

PROPOSITION 5. *Let E be the inductive limit of the convex spaces E_γ by the mappings u_γ, and let t be a linear mapping of E into a convex space F. Then t is continuous if and only if, for each γ, $t \circ u_\gamma$ is a continuous mapping of E_γ into F. More generally, the set T of* *linear mappings of E into F is equicontinuous if and only if each set $T \circ u_\gamma$ is equicontinuous.*

Proof. The mapping t is continuous if and only if, for each absolutely convex neighbourhood V in F, $t^{-1}(V)$ is a neighbourhood in E. By Proposition 4, we require, for each γ,

$$u_\gamma^{-1}(t^{-1}(V)) = (t \circ u_\gamma)^{-1}(V)$$

to be a neighbourhood in E_γ, i.e. we require the continuity of each $t \circ u_\gamma$.

For the equicontinuity of T, we require that $\bigcap_{t \in T} t^{-1}(V)$ be a neighbourhood in E, and a similar argument shows that this is equivalent to the equicontinuity of each $T \circ u_\gamma$.

COROLLARY. *A separated inductive limit topology is the topology of uniform convergence on the subsets A' of the dual such that, for each γ, $u'_\gamma(A')$ is equicontinuous, u'_γ being the transpose of u_γ.*

For A' is equicontinuous if and only if each $u'_\gamma(A') = A' \circ u_\gamma$ is equicontinuous.

An extreme case of an inductive limit topology is the quotient topology studied in § 1. For if $E = E_0/M$ and k is the canonical mapping of E_0 onto E, the quotient topology on E is the finest convex topology making k continuous.

Very often the convex spaces E_γ are vector subspaces of E whose union spans E, and the linear mappings u_γ are all (restrictions to E_γ of) the identity mapping of E. Then the inductive limit topology is the finest convex topology on E which induces on each E_γ a topology coarser than the given topology, and an absolutely convex set U is a neighbourhood in E if and only if $U \cap E_\gamma$ is a neighbourhood in E_γ for each γ. Finally, if t is a linear mapping of E into another convex space, t is continuous if and only if it is continuous on each E_γ, and there is a similar criterion for equicontinuity.

A general inductive limit (of the convex spaces E_γ by the mappings u_γ, say) can be reduced to this special case. For $u_\gamma(E_\gamma)$ is a vector subspace of E and the topology of E_γ can be transferred to $u_\gamma(E_\gamma)$ by taking for the neighbourhoods in $u_\gamma(E_\gamma)$ the images by u_γ of the neighbourhoods in E_γ. Then it is easy to see, from the characterisation of the neighbourhoods in E given in

Proposition 4, that E is also the inductive limit of its vector subspaces $u_\gamma(E_\gamma)$.

Certain properties of convex spaces are preserved by the operation of taking inductive limits.

PROPOSITION 6. *An inductive limit of barrelled spaces is barrelled.*

Proof. Let B be a barrel in E, the inductive limit of the barrelled spaces E_γ by the mappings u_γ. Then, for each γ, $u_\gamma^{-1}(B)$ is closed and is therefore a barrel in E_γ. Hence $u_\gamma^{-1}(B)$ is a neighbourhood in E_γ, and so, by Proposition 4, B is a neighbourhood in E.

COROLLARY. *A quotient of a barrelled space is barrelled.*

We shall discuss in the next section another property which is preserved under inductive limits. It is perhaps worth mentioning that an inductive limit of separated convex spaces need not be separated (but see Ch. VII, Prop. 2).

It is possible to define an inductive limit topology on a vector space E by the mappings u_γ from E_γ into E without assuming that $\bigcup_{\gamma \in \Gamma} u_\gamma(E_\gamma)$ spans E. Then the neighbourhoods in E are those absolutely convex absorbent sets U for which $u_\gamma^{-1}(U)$ is a neighbourhood in E_γ for each γ. With this one modification, the results of this section remain valid for these more general inductive limits.

3. Mackey spaces. Suppose that E is a convex space and that t is a linear mapping of E into another convex space F. We have already seen (Ch. III, Prop. 1) that if t is continuous, then t maps bounded subsets of E into bounded subsets of F. Let us (temporarily) call t *bounded* if it has this property. We shall prove that if E is normed, or metrisable, then conversely the boundedness of t ensures its continuity. A convex space E with the property that every bounded linear mapping on it is continuous will be called a *Mackey space*. (It is sometimes called *bornological.*) A separated Mackey space E with dual E' has the Mackey topology $\tau(E, E')$, for if ξ is the topology of E, the identity mapping of E under ξ onto E under $\tau(E, E')$ is bounded (Ch. IV, Th. 1) and so continuous. (Some authors have used the term 'Mackey space' for any convex space with the Mackey topology, but we reserve it for this more restrictive class.)

PROPOSITION 7. *An inductive limit of Mackey spaces is a Mackey space.*

Proof. Let E be the inductive limit of the Mackey spaces E_γ by the mappings u_γ and let t be a bounded linear mapping of E into the convex space F. If A is a bounded subset of E_γ, the continuity of u_γ and the boundedness of t ensure that $(t \circ u_\gamma)(A) = t(u_\gamma(A))$ is bounded in F. Thus $t \circ u_\gamma$ is bounded and so continuous. Hence t is continuous (Prop. 5).

PROPOSITION 8. *Every metrisable convex space is a Mackey space.*

Proof. Let E be metrisable and let (U_n) be a base of neighbourhoods, with $U_{n+1} \subseteq U_n$ for all n. If the bounded linear mapping t of E into the convex space F is not continuous, there is some neighbourhood V in F for which $t^{-1}(V)$ is not a neighbourhood in E. Hence there are points x_n in U_n with $t(x_n) \notin nV$. Then $\{x_n\}$ is clearly a bounded set, whereas $\{t(x_n)\}$ is not. This contradiction establishes the result.

COROLLARY. *Every inductive limit of metrisable convex spaces is a Mackey space.*

Leading to a form of converse of this Corollary we have:

LEMMA 1. *Let E be a separated convex space with topology ξ. Then there is a finest convex topology η on E under which E has the same bounded sets as under ξ. Under the topology η, E is a Mackey space, the inductive limit of a family of normed vector subspaces spanning E. The topologies ξ and η are identical if and only if E is a Mackey space under ξ.*

Proof. Let A be any absolutely convex subset of E that is closed and bounded in the topology ξ, and let E_A be the vector subspace spanned by A. Because A is bounded and E separated under ξ, the gauge of A is a norm on E_A, making it a normed space with topology, η_A say, finer than the topology induced on E_A by ξ. Now let $\mathscr{U}$ be a base of neighbourhoods in a convex topology ζ on E. The set A is bounded for ζ if and only if, for each $U \in \mathscr{U}$, there is a $\lambda > 0$ with $\lambda A \subseteq U \cap E_A$, i.e. if and only if ζ induces on E_A a topology coarser than η_A. Now every ξ-bounded set is contained in a ξ-closed absolutely convex ξ-bounded set; thus the finest convex topology ζ for which every ξ-bounded set is ζ-bounded is the inductive limit topology η of the normed spaces E_A. Since η is finer than ξ, every η-bounded set is ξ-bounded, and so η is the finest convex topology for which the same sets are bounded as for ξ. By the Corollary of Proposition 8, E is a Mackey space under η. Finally, if E is a Mackey space under ξ, the identity mapping of E under ξ into E under η is bounded and therefore continuous. Hence ξ is finer than η and so they are identical.

LEMMA 2. *If, in Lemma 1, E is complete under ξ, then under η, E is the inductive limit of a family of Banach spaces.*

Proof. We show that each E_A is complete under η_A. If (x_n) is any Cauchy sequence in E_A, then (x_n) is also ξ-Cauchy and so converges to some point a of E under ξ. We prove that $a \in E_A$ and that $x_n \to a$ under η_A. For each $\epsilon > 0$ there is an n_0 with $x_m - x_n \in \epsilon A$ for $m, n \geqslant n_0$. The set ϵA is closed in E under ξ and so, letting $m \to \infty$, we have $a - x_n \in \epsilon A$ for $n \geqslant n_0$. This shows that $a \in E_A$ and that $x_n \to a$ in E_A. Thus E_A is complete.

In the proof of Lemma 2 it is sufficient to know that every ξ-Cauchy sequence in E is convergent, a condition weaker than completeness if E is not metrisable.

THEOREM 1. *A separated convex space is a Mackey space if and only if it is an inductive limit of normed spaces. A complete separated Mackey space is an inductive limit of Banach spaces.*

Proof. The Corollary of Proposition 8 and Lemmas 1 and 2 give the results.

It is easy to extend Theorem 1 to non-separated spaces by replacing each normed space by a space whose topology is given by a single seminorm.

PROPOSITION 9. *In the dual E' of a separated Mackey space E, every $\beta(E', E)$-bounded subset is equicontinuous.*

Proof. The topology on E of uniform convergence on the $\beta(E', E)$-bounded subsets of E' is finer than the initial topology on E and has the same bounded sets, and so (Lemma 1) these topologies coincide.

COROLLARY. *The identity mapping of a separated Mackey space into its bidual is an isomorphism* (cf. Ch. IV, Prop. 2).

PROPOSITION 10. *A complete separated Mackey space is barrelled.*

Proof. By Lemma 2, a complete separated Mackey space is an inductive limit of Banach spaces. Each of these is barrelled (Ch. IV, Th. 2) and so is their inductive limit (Prop. 6).

We shall prove later (Ch. VI, Prop. 1) that the dual E' of a separated Mackey space is complete under various topologies, including $\beta(E', E)$.

4. Projective limits. The second general method of topologising a vector space is, in a sense which we shall make precise later, dual to that of taking inductive limits. This time we topologise a vector space F by means of linear mappings v_γ of F into convex spaces F_γ, such that, if $x \in F$ and $x \neq o$, there is some γ for which $v_\gamma(x) \neq o$.

PROPOSITION 11. *Let F be a vector space, and for each $\gamma \in \Gamma$ let v_γ be a linear mapping of F into a convex space F_γ, such that*

$$\bigcap_{\gamma \in \Gamma} v_\gamma^{-1}(o) = \{o\}.$$

Then there is a coarsest topology on F compatible with the algebraic structure under which all the v_γ are continuous; under this topology F is a convex space. If $\mathscr{V}_\gamma$ is a base of absolutely convex neighbourhoods in F_γ, the finite intersections of the sets $v_\gamma^{-1}(V_\gamma)$ $(V_\gamma \in \mathscr{V}_\gamma, \gamma \in \Gamma)$, form a base $\mathscr{V}$ of absolutely convex neighbourhoods for F with this topology.

Proof. If the v_γ are to be continuous, the sets of $\mathscr{V}$ must be neighbourhoods in F. But these form a base for a convex topology on F (Ch. I, Th. 2), which is the coarsest making the v_γ continuous.

The convex space F with this topology is called the *projective limit* of the convex spaces F_γ by the mappings v_γ.

PROPOSITION 12. *Let E be a convex space and t a linear mapping of E into the projective limit F of the convex spaces F_γ by the mappings v_γ. Then t is continuous if and only if for each γ, $v_\gamma \circ t$ is a continuous mapping of E into F_γ.*

Proof. Clearly t is continuous if and only if, for each γ and each $V_\gamma \in \mathscr{V}_\gamma$, $t^{-1}(v_\gamma^{-1}(V_\gamma)) = (v_\gamma \circ t)^{-1}(V_\gamma)$ is a neighbourhood in E, and this is the condition that each $v_\gamma \circ t$ should be continuous.

Proposition 13. *A projective limit of separated spaces is separated.*

Proof. With the notation of Proposition 11,

$$\bigcap_{V \in \mathscr{V}} V = \bigcap_{\gamma \in \Gamma} \bigcap_{V_\gamma \in \mathscr{V}_\gamma} v_\gamma^{-1}(V_\gamma) = \bigcap_{\gamma \in \Gamma} v_\gamma^{-1}(o) = \{o\}.$$

It is only in special cases that we can give a simple characterisation of bounded or precompact sets in an inductive limit. In a projective limit, however, we have:

Proposition 14. *Let F be the projective limit of the convex spaces F_γ by the mappings v_γ. Then the subset A of F is bounded, or precompact, if and only if each $v_\gamma(A)$ has the same property.*

Proof. If A is bounded, or precompact, then so is each $v_\gamma(A)$ since v_γ is continuous. If each $v_\gamma(A)$ is bounded and if V_γ is an absolutely convex neighbourhood in F_γ, then V_γ absorbs $v_\gamma(A)$ and so $v_\gamma^{-1}(V_\gamma)$ absorbs A. This, by Proposition 11, ensures that A is bounded in F. The proof for precompactness is similar.

One of the most immediate examples of a projective limit is the weak topology on any convex space F, obtained by taking for (v_γ) the set of all continuous linear forms on F (so that each F_γ is the scalar field). An extreme example of a projective limit is the induced topology on a vector subspace M of a convex space F; it is the coarsest topology making the identity mapping of M into F continuous. We shall consider another example in the next section.

The dual of an inductive limit is a projective limit. More precisely:

Proposition 15. *For each $\gamma \in \Gamma$, let E_γ be a separated convex space and let its dual E'_γ have the topology of $\mathscr{A}_\gamma$-convergence. Let E be the inductive limit of the spaces E_γ by the mappings u_γ, and suppose that E is separated. Then, if $\mathscr{A}$ is the set of finite unions of the sets of $\{u_\gamma(\mathscr{A}_\gamma): \gamma \in \Gamma\}$, the dual E' of E under the topology of $\mathscr{A}$-convergence is the projective limit of the E'_γ by the mappings u'_γ, where u'_γ is the transpose of u_γ.*

Proof. Each u'_γ maps E' into E'_γ (Ch. II, Props. 13 and 12). Also

$$\bigcap_{\gamma \in \Gamma} u_\gamma'^{-1}(o) = \{o\},$$

since if $u'_\gamma(x') = o$ for all γ,

$$\langle u_\gamma(x_\gamma), x' \rangle = \langle x_\gamma, u'_\gamma(x') \rangle = 0$$

for all $x_\gamma \in E_\gamma$ and all γ; hence x' vanishes on $\bigcup_{\gamma \in \Gamma} u_\gamma(E_\gamma)$, which spans E. Now the projective limit topology on E' is the coarsest in which the sets $u'^{-1}_\gamma(A^0_\gamma)\,(A_\gamma \in \mathscr{A}_\gamma)$ are neighbourhoods (Prop. 11). Since

$$u'^{-1}_\gamma(A^0_\gamma) = (u_\gamma(A_\gamma))^0$$

(Ch. II, Lemma 6) this topology is that of $\mathscr{A}$-convergence.

COROLLARY. *Let E be a separated convex space with dual E', and let M be a closed vector subspace of E. Then the topology of $\mathscr{A}$-convergence on E' induces on the dual M^0 of E/M the topology of $k(\mathscr{A})$-convergence, where k is the canonical mapping of E onto E/M.*

Proof. The transpose mapping k' is the identity mapping of M^0 into E' and so the induced topology on M^0 is the projective limit of E' by k'. By the proposition, this is the topology of $k(\mathscr{A})$-convergence, since E/M is the inductive limit of E by the mapping k.

Only a Mackey space can be an inductive limit of normed spaces. On the other hand:

PROPOSITION 16. *Every separated convex space is a projective limit of normed spaces.*

Proof. For each continuous seminorm p on the convex space E, $p^{-1}(0)$ is a vector subspace and p defines a norm on $E_p = E/p^{-1}(0)$ (for each $X \in E_p$, take $\|X\|$ to be the common value of $p(x)$ for $x \in X$). Then E is clearly the projective limit of the E_p by the canonical mappings k_p of E onto E_p, the property $\bigcap k_p^{-1}(o) = \{o\}$ being equivalent to the separatedness of E.

Since every normed space can be embedded in a Banach space (this will be proved in Ch. VI, § 1) it follows that every separated convex space is a projective limit of Banach spaces.

As with inductive limits, the special condition

$$\bigcap_{\gamma \in \Gamma} v_\gamma^{-1}(o) = \{o\}$$

imposed in the definition of projective limits may be dispensed with for some applications. Except for Proposition 13, which is no longer valid, the results of this section all hold. Moreover, every convex space (separated or not) is a projective limit of normed or Banach spaces in this more general sense. The dual of an inductive limit in the wider sense mentioned at the end of § 2 is a projective limit in the wider sense.

5. Product spaces. If $(E_\gamma: \gamma \in \Gamma)$ is any family of sets, their (Cartesian) product $\times E_\gamma$ (or, more formally, $\underset{\gamma \in \Gamma}{\times} E_\gamma$) is the set of of all families $x = (x_\gamma)$ with $x_\gamma \in E_\gamma$ for each $\gamma \in \Gamma$. By analogy with the more familiar case in which Γ is a finite set, the elements x_γ are called the *coordinates* of x, and the mapping p_γ which assigns to each x its γ-coordinate x_γ is called the *projection mapping* of $\times E_\gamma$ onto E_γ.

When the E_γ are all vector spaces over the same field Φ, $\times E_\gamma$ can be made into a vector space over Φ by defining $(x_\gamma)+(y_\gamma)$ to be $(x_\gamma + y_\gamma)$ and $\lambda(x_\gamma)$ to be (λx_γ). The projection p_γ is then a linear mapping of $\times E_\gamma$ onto E_γ.

If each E_γ is a convex space, the space $E = \times E_\gamma$ can be made into a convex space by regarding it as the projective limit of the spaces E_γ by the mappings p_γ. This topology is called the *product topology* on E; it is the coarsest under which the projection mappings p_γ are continuous. If, for each γ, $\mathscr{U}_\gamma$ is a base of absolutely convex neighbourhoods in E_γ, the finite intersections of the sets $p_\gamma^{-1}(U_\gamma)$ $(U_\gamma \in \mathscr{U}_\gamma, \gamma \in \Gamma)$ form a base of absolutely convex neighbourhoods in E (Prop. 11). A linear mapping t of any convex space into E is continuous if and only if each $p_\gamma \circ t$ is continuous (Prop. 12). A subset A of E is bounded, or precompact, if and only if each of its projections $p_\gamma(A)$ has the same property (Prop. 14). The conclusion of Proposition 13 can be strengthened slightly for product spaces:

Proposition 17. *The product $\times E_\gamma$ is separated if and only if each E_γ is separated.*

Proof. The point x is in every neighbourhood in the product if and only if each $p_\gamma(x)$ is in every neighbourhood in E_γ.

Proposition 18. *The subset $A = \times A_\gamma$ of $\times E_\gamma$ is complete if and only if each of its projections A_γ is complete.*

Proof. If each A_γ is complete and $\mathscr{F}$ is a Cauchy filter on E with $A \in \mathscr{F}$, then $p_\gamma(\mathscr{F})$ is Cauchy and $A_\gamma \in p_\gamma(\mathscr{F})$; hence $p_\gamma(\mathscr{F})$ converges to $x_\gamma \in A_\gamma$. Then $\mathscr{F}$ converges to $x = (x_\gamma) \in A$. Conversely, if A is complete and $\mathscr{G}$ is a Cauchy filter on some E_γ with $A_\gamma \in \mathscr{G}$, then there exists a Cauchy filter $\mathscr{F}$ on E with $A \in \mathscr{F}$ and $p_\gamma(\mathscr{F}) = \mathscr{G}$. For, if $\delta \neq \gamma$, choose any Cauchy filter $\mathscr{F}_\delta$ on E_δ, with $A_\delta \in \mathscr{F}_\delta$; the products of the sets of $\mathscr{G}$ and the $\mathscr{F}_\delta$ generate such a filter $\mathscr{F}$. Now $\mathscr{F}$ converges, to $x \in A$, and so $\mathscr{G}$ converges to $p_\gamma(x) \in A_\gamma$.

Corollary 1. *The product* $\mathsf{X}E_\gamma$ *is complete if and only if each* E_γ *is complete.*

Corollary 2. *The subset* $A = \mathsf{X}A_\gamma$ *of* $\mathsf{X}E_\gamma$ *is compact if and only if each of its projections* A_γ *is compact.*

We now have two special cases of projective limits, the induced topology on a vector subspace and the product topology. These are in a sense the only cases:

Proposition 19. *Let F be the projective limit of the convex spaces F_γ ($\gamma \in \Gamma$) by the mappings v_γ. Then F is isomorphic to a vector subspace of the product* $\mathsf{X}_{\gamma \in \Gamma} F_\gamma$.

Proof. We define a linear mapping v of F into $\mathsf{X}F_\gamma$ by putting $v(x) = (v_\gamma(x))$; then since $p_\gamma \circ v = v_\gamma$ for each γ and each v_γ is continuous, v is also continuous. Also

$$v^{-1}(o) = \bigcap_{\gamma \in \Gamma} v_\gamma^{-1}(o) = \{o\},$$

by the special assumption made for projective limits, and so v is a (1, 1) mapping of F onto $v(F)$. If v^{-1} is its inverse on $v(F)$, $v_\gamma \circ v^{-1} = p_\gamma$ for each γ and so (Prop. 12) v^{-1} is also continuous. Thus v is an isomorphism of F onto $v(F)$.

Corollary. *Every separated convex space is isomorphic to a vector subspace of a product of normed* (*or Banach*) *spaces.*

If E is the product of the vector spaces E_γ, there is a natural way of embedding each E_γ in E. For each $x_\gamma \in E_\gamma$, denote by $j_\gamma(x_\gamma)$ the element of E all of whose coordinates are o except the γth which is x_γ; then j_γ is a (1, 1) linear mapping, called the *injection mapping* of E_γ into $E = \mathsf{X}E_\gamma$. Algebraically, $p_\gamma \circ j_\gamma$ is the identity mapping on E_γ, and, in fact, p_γ restricted to

$j_\gamma(E_\gamma)$ is the inverse of j_γ. If γ and δ are distinct members of Γ, $p_\delta \circ j_\gamma$ is the zero mapping of E_γ into E_δ.

PROPOSITION 20. *If each E_γ is a convex space and $\underset{\gamma \in \Gamma}{\mathsf{X}} E_\gamma$ has the product topology, then each injection j_γ is an isomorphism of E_γ onto $j_\gamma(E_\gamma)$. If each E_γ is separated, then $j_\gamma(E_\gamma)$ is closed in $\underset{\gamma \in \Gamma}{\mathsf{X}} E_\gamma$.*

Proof. First, j_γ is continuous because $p_\delta \circ j_\gamma$ is continuous for each $\delta \in \Gamma$. Next, $j_\gamma^{-1} = p_\gamma$ (restricted to $j_\gamma(E_\gamma)$) and so j_γ^{-1} is also continuous. Thus j_γ is an isomorphism. Finally,

$$j_\gamma(E_\gamma) = \bigcap_{\delta \in \Gamma,\, \delta \neq \gamma} p_\delta^{-1}(o);$$

if E_δ is separated then $\{o\}$ is a closed set in each E_δ, and so each $p_\delta^{-1}(o)$, and their intersection, is closed.

Because of this result, it is often convenient to identify E_γ with its isomorphic image $j_\gamma(E_\gamma)$ in $\underset{\gamma \in \Gamma}{\mathsf{X}} E_\gamma$. When this is done the convex spaces E_γ become vector subspaces of $\underset{\gamma \in \Gamma}{\mathsf{X}} E_\gamma$, disjoint except for the origin. In an entirely similar way, if $\Delta \subseteq \Gamma$, the topological product $\underset{\gamma \in \Delta}{\mathsf{X}} E_\gamma$ can be isomorphically embedded in $\underset{\gamma \in \Gamma}{\mathsf{X}} E_\gamma$, in which it is a closed subspace provided that each E_γ is separated.

6. Direct sums. In the product $\mathsf{X} E_\gamma$ of the vector spaces E_γ, the vector subspace spanned by $\bigcup E_\gamma$ (or, more precisely, by $\bigcup j_\gamma(E_\gamma)$, where j_γ is the injection mapping of E_γ into the product) is called the *direct sum* of the vector spaces E_γ, and denoted by ΣE_γ (or, formally, by $\sum_{\gamma \in \Gamma} E_\gamma$). It is the set of those elements of $\mathsf{X} E_\gamma$ with only a finite number of non-zero coordinates. Clearly ΣE_γ is the whole of $\mathsf{X} E_\gamma$ whenever Γ is a finite set. If p_γ is the projection mapping of $\mathsf{X} E_\gamma$ onto E_γ, it is clear that any element x of ΣE_γ (other than the origin) is the sum of the finite number of its non-zero coordinates $p_\gamma(x)$. It is natural to write $x = \Sigma p_\gamma(x)$, interpreting this last sum as meaning the sum of its non-zero terms. If $x = o$, all the $p_\gamma(x)$ are zero and the expression $x = \Sigma p_\gamma(x)$ still holds under the reasonable interpretation of the sum as zero.

If each E_γ is a convex space, the direct sum $E = \Sigma E_\gamma$ can be given the topology induced on it by the product topology for $\times E_\gamma$; we shall call this the product topology on ΣE_γ. We have already seen (Prop. 20) that this topology induces the original topology on each E_γ, but in general it is not the finest convex space topology to do so. The finest such topology is obtained by considering E as the inductive limit of the convex spaces E_γ by the injection mappings j_γ. This topology is called the *direct sum topology* for E, and under it E is called the *topological direct sum* of the convex spaces E_γ.

PROPOSITION 21. *The direct sum topology on ΣE_γ is finer than the product topology. For every finite subset Δ of Γ, these two topologies coincide on $\sum_{\gamma \in \Delta} E_\gamma$ (regarded as a vector subspace of $\sum_{\gamma \in \Gamma} E_\gamma$). The direct sum topology induces the original topology on each E_γ.*

Proof. Let ξ be the direct sum topology and η the product topology. Each injection j_γ is continuous in η, and so η must be coarser than ξ, which is the finest making the j_γ continuous.

If Δ has n elements and U is any absolutely convex ξ-neighbourhood, then $U \cap E_\gamma$ is a neighbourhood in E_γ for each γ. Hence

$$V = n^{-1} \bigcap_{\gamma \in \Delta} p_\gamma^{-1}(U \cap E_\gamma)$$

is an η-neighbourhood. But on $\sum_{\gamma \in \Delta} E_\gamma$, if $x \in V$ then $p_\gamma(x) \in n^{-1}U$ for each $\gamma \in \Delta$ and so

$$x = \sum_{\gamma \in \Delta} p_\gamma(x) \in U.$$

Hence
$$V \cap \sum_{\gamma \in \Delta} E_\gamma \subseteq U \cap \sum_{\gamma \in \Delta} E_\gamma,$$

so that η is also finer than ξ on $\sum_{\gamma \in \Delta} E_\gamma$. In the special case $\Delta = \{\gamma\}$, Proposition 20 gives the last part.

In general the product and direct sum topologies coincide only on finite direct sums: if Δ is infinite and each E_γ with $\gamma \in \Delta$ contains a neighbourhood U_γ other than E_γ itself, the direct sum topology is strictly finer than the product topology on $\sum_{\gamma \in \Delta} E_\gamma$, because $\bigcap_{\gamma \in \Delta} p_\gamma^{-1}(U_\gamma)$ is a neighbourhood in one topology but not the other.

Proposition 22. *The topological direct sum* ΣE_γ *is separated if and only if each* E_γ *is separated. When it is, each* E_γ *is closed in* ΣE_γ.

Proof. If ΣE_γ is separated, so is each E_γ, because the direct sum topology ξ induces on it the original topology (Prop. 21). If each E_γ is separated, so is $\mathsf{X}E_\gamma$ under the product topology η (Prop. 17) and therefore ΣE_γ is separated under the finer topology ξ. Then, since E_γ is closed in $\mathsf{X}E_\gamma$ under η (Prop. 20), E_γ is closed in ΣE_γ under ξ.

Our next result deals with completeness in topological direct sums. We lead up to it by:

Lemma 3. *The topological direct sum* ΣE_γ *has a base of absolutely convex neighbourhoods which are closed in the product topology.*

Proof. For each $\gamma \in \Gamma$, let $\mathscr{V}_\gamma$ be a base of absolutely convex neighbourhoods in E_γ. Then (Prop. 4, Corollary) the absolutely convex envelopes of the sets of the form $\bigcup V_\gamma$ $(V_\gamma \in \mathscr{V}_\gamma)$ form a neighbourhood base $\mathscr{V}$ for the direct sum topology ξ on $E = \Sigma E_\gamma$. We prove that the closures of the sets of $\mathscr{V}$ in the product topology η form the required neighbourhood base, by showing that, for each $V \in \mathscr{V}$, its η-closure is contained in $3V$.

Let x be in the η-closure of V. Then there is a finite subset Δ of Γ (containing, say, n elements) such that $p_\gamma(x) = o$ for $\gamma \notin \Delta$. Since

$$W = n^{-1} \bigcap_{\gamma \in \Delta} p_\gamma^{-1}(V \cap E_\gamma)$$

is an η-neighbourhood, there is a point $y \in V$ with $x - y \in W$. Hence $p_\gamma(x-y) \in n^{-1}V$ for $\gamma \in \Delta$ and therefore

$$\sum_{\gamma \in \Delta} p_\gamma(x-y) \in V.$$

Since $y \in V$, $y = \Sigma \lambda_\gamma y_\gamma$, where $\Sigma |\lambda_\gamma| \leqslant 1$ (only a finite number of the λ_γ being non-zero) and $y_\gamma \in V_\gamma \subseteq V \cap E_\gamma$. Thus

$$p_\gamma(y) = \lambda_\gamma y_\gamma \in \lambda_\gamma V.$$

Hence

$$\sum_{\gamma \in \Gamma \cap (\sim\Delta)} p_\gamma(x-y) = -\sum_{\gamma \in \Gamma \cap (\sim\Delta)} p_\gamma(y) \subseteq \Big(\sum_{\gamma \in \Gamma \cap (\sim\Delta)} |\lambda_\gamma|\Big) V \subseteq V.$$

But

$$\sum_{\gamma \in \Delta} p_\gamma(x-y) \in V \quad \text{and so} \quad x - y = \sum_{\gamma \in \Gamma} p_\gamma(x-y) \in 2V;$$

thus $x \in 3V$.

PROPOSITION 23. *The topological direct sum of the separated convex spaces E_γ is complete if and only if each E_γ is complete.*

Proof. If $E = \Sigma E_\gamma$ is complete, so are its closed vector subspaces E_γ. Suppose, then, that each E_γ is complete and let $\mathscr{F}$ be a Cauchy filter on E. Then $\mathscr{F}$ is also the base of a Cauchy filter on $\mathsf{X} E_\gamma$ under the product topology η, which is complete (Prop. 18, Cor. 1). Hence $\mathscr{F}$ converges under η to a point x of $\mathsf{X} E_\gamma$. We show first that $x \in E$.

Let Δ be the subset of Γ for which $p_\gamma(x) \neq o$. Then, for each $\gamma \in \Delta$, there is an absolutely convex neighbourhood U_γ in E_γ with $p_\gamma(x) \notin U_\gamma$. Since

$$U = \tfrac{1}{2} \bigcap_{\gamma \in \Delta} p_\gamma^{-1}(U_\gamma)$$

is a neighbourhood in the direct sum topology ξ on E, $\mathscr{F}$ has a set A small of order U. Take any point $y \in A$. If Δ is infinite, there is a $\delta \in \Delta$ with $p_\delta(y) = o$. Since x lies in the η-closure of A, the η-neighbourhood $x + \frac{1}{2} p_\delta^{-1}(U_\delta)$ of x meets A and so

$$p_\delta(x) \in p_\delta(y) + p_\delta(U) + \tfrac{1}{2} p_\delta(p_\delta^{-1}(U_\delta)) \subseteq U_\delta.$$

This is a contradiction and so Δ must be finite and $x \in E$.

Finally we prove that $\mathscr{F} \to x$ in ξ. Let $\mathscr{V}$ be the base of ξ-neighbourhoods closed in η whose existence is established in Lemma 3, and let $V \in \mathscr{V}$. Then $\mathscr{F}$ contains a set B small of order $\frac{1}{2}V$; if z is any point of B, we have $B \subseteq z + \frac{1}{2}V$. Since x belongs to the η-closure of B and V is η-closed, $x \in z + \frac{1}{2}V$. Hence $z \in x + \frac{1}{2}V$ and $B \subseteq x + \frac{1}{2}V + \frac{1}{2}V = x + V$. Thus $\mathscr{F} \to x$ in ξ and E is complete.

PROPOSITION 24. *In a separated topological direct sum ΣE_γ, the set A is bounded, or precompact, if and only if it is contained in a finite sum of subsets of the E_γ with the same property.*

Proof. Any set of such a form is clearly bounded, or precompact. If A is a bounded, or precompact, set, then each of its projections $p_\gamma(A)$ has the same property because of the continuity of each p_γ. It is therefore enough to show that at most a finite number of these projections are different from $\{o\}$, since $A \subseteq \Sigma p_\gamma(A)$. If not, there is a sequence $(\gamma(n))$ and a sequence of points (x_n) with $x_n \neq o$, $x_n \in p_{\gamma(n)}(A)$. There are then absolutely convex neighbourhoods $U_{\gamma(n)}$ with $x_n \notin nU_{\gamma(n)}$. For $\gamma \notin \{\gamma(n)\}$ take $U_\gamma = E_\gamma$; then the convex envelope of these neighbourhoods is a neighbourhood U in the direct sum topology, with $p_{\gamma(n)}(U) \subseteq U_{\gamma(n)}$ for each n. Thus $A \nsubseteq nU$ for any n, and this contradicts the boundedness of A.

Corollary. *In a separated topological direct sum ΣE_γ, the closed set A is compact if and only if it is contained in a finite sum of compact subsets of the E_γ.*

For such a set is compact (Ch. III, Lemmas 7 (ii) and 6 (ii)). If A is compact, the proof of the proposition shows that A is a closed subset of a finite sum of projections $p_\gamma(A)$, and these are compact (Ch. III, Prop. 7).

Proposition 25. *The dual of the topological direct sum ΣE_γ is the product $\mathsf{X} E'_\gamma$ of the duals. If each E_γ is separated, and if each E'_γ has the topology of $\mathscr{A}_\gamma$-convergence, then the product topology on $\mathsf{X} E'_\gamma$ is the topology of $\mathscr{A}$-convergence, where $\mathscr{A}$ is the set of all finite unions of sets of $\bigcup \mathscr{A}_\gamma$.*

Proof. A linear form x' on $E = \Sigma E_\gamma$ is continuous if and only if each of its restrictions x'_γ to E_γ is continuous, and the identity $\langle x, x' \rangle = \Sigma \langle p_\gamma(x), x'_\gamma \rangle$ establishes an algebraic isomorphism between the dual of E and $\mathsf{X} E'_\gamma$. The second part follows from Proposition 15, the transpose of j_γ being the projection p'_γ of $\mathsf{X} E'_\gamma$ onto E'_γ.

Unlike Proposition 15, this proposition has a dual form in which the roles of direct sum and product are interchanged.

Proposition 26. *The dual of the topological product $\mathsf{X} E_\gamma$ is the direct sum $\Sigma E'_\gamma$ of the duals. If each E_γ is separated and E'_γ has the topology of $\mathscr{A}_\gamma$-convergence (where the sets of $\mathscr{A}_\gamma$ are supposed to be absolutely convex), then the direct sum topology on $\Sigma E'_\gamma$ is the topology of $\mathscr{A}$-convergence, where $\mathscr{A}$ is the set of all products $\mathsf{X} A_\gamma$ with $A_\gamma \in \mathscr{A}_\gamma$ for each $\gamma \in \Gamma$.*

Proof. Let x' be a continuous linear form on the product, bounded on the neighbourhood

$$U = \bigcap_{\gamma \in \Delta} p_\gamma^{-1}(U_\gamma),$$

where Δ is a finite subset of Γ and each U_γ is a neighbourhood in E_γ. Then x' vanishes on E_γ for each $\gamma \notin \Delta$, and so $x' = \sum_{\gamma \in \Delta} x'_\gamma$, where $x'_\gamma = x' \circ j_\gamma$ is the restriction of x' to E_γ. Thus $x' \in \Sigma E'_\gamma$.

Now the direct sum topology has a base of neighbourhoods consisting of the absolutely convex envelopes of sets of the form $\bigcup A^0_\gamma$, with $A_\gamma \in \mathscr{A}_\gamma$ (the polar of A_γ being taken in E'_γ). We have to show that the sets A^0 ($A \in \mathscr{A}$) define the same topology. The proof will therefore be complete if we show that, for each $A = \mathsf{X} A_\gamma \in \mathscr{A}$, $V' \subseteq A^0 \subseteq 2V'$, where V' is the absolutely convex envelope of $\bigcup A^0_\gamma$.

The first relation $V' \subseteq A^0$ is straightforward. To prove the second, let $x' = \Sigma x'_\gamma \in A^0$ and denote by Δ the finite subset of Γ for which

$$\lambda_\gamma = \sup_{x_\gamma \in A_\gamma} |\langle x_\gamma, x'_\gamma \rangle| > 0.$$

Then $x' - \sum_{\gamma \in \Delta} x'_\gamma$ vanishes on A and so certainly belongs to V'; also

$$\sum_{\gamma \in \Delta} x'_\gamma = \sum_{\gamma \in \Delta} \lambda_\gamma (x'_\gamma / \lambda_\gamma) \in \sum_{\gamma \in \Delta} \lambda_\gamma A^0_\gamma \subseteq V',$$

since

$$\sup_{x \in A} |\langle x, x' \rangle| = \sum_{\gamma \in \Delta} \lambda_\gamma \leqslant 1.$$

Hence $x' \in V' + V' = 2V'$.

We can use this proposition to help us enlarge the class of barrelled spaces:

Proposition 27. *A product of separated barrelled spaces is barrelled.*

Proof. Let $E = \mathsf{X} E_\gamma$ be the product and $E' = \Sigma E'_\gamma$ its dual (Prop. 26). On each E'_γ choose any topology of the dual pair (E'_γ, E_γ), e.g. $\sigma(E'_\gamma, E_\gamma)$; then the direct sum topology ξ' on E' has dual E (Prop. 25). Hence if A' is weakly bounded, it is also ξ'-bounded (Ch. IV, Th. 1) and so (Prop. 24) $A' \subseteq \sum_{\gamma \in \Delta} A'_\gamma$, where Δ is finite and each A'_γ is a weakly bounded set in E'_γ. But each E_γ is barrelled and therefore each A'_γ is equicontinuous; hence A' is equicontinuous and thus E is barrelled.

We now have two special cases of inductive limits, the quotient topology and the direct sum topology. These are again the only cases:

Proposition 28. *Let E be the inductive limit of the convex spaces E_γ by the mappings u_γ. Then E is isomorphic to a quotient of the direct sum ΣE_γ.*

Proof. We define a linear mapping u of ΣE_γ onto E by putting $u(\Sigma x_\gamma) = \Sigma u_\gamma(x_\gamma)$, the special condition that $\bigcup u_\gamma(E_\gamma)$ spans E ensuring that u maps the direct sum onto E. Since $u \circ j_\gamma = u_\gamma$ for each γ and each u_γ is continuous, u is also continuous. Now u can be decomposed into $v \circ k$, where k is the canonical mapping of ΣE_γ onto $F = \Sigma E_\gamma / u^{-1}(o)$ and v is a continuous $(1, 1)$ mapping of F onto E (Prop. 2). We prove v^{-1} continuous. For each γ, $v^{-1} \circ u_\gamma = k \circ j_\gamma$, which is continuous; hence (Prop. 5) v^{-1} is continuous. Thus v is an isomorphism of F onto E.

7. Topological supplements. Suppose that E is a convex space and that M and N are vector subspaces of E such that E is the (algebraic) direct sum of M and N. (Necessary and sufficient conditions for this are that $M + N = E$ and $M \cap N = \{o\}$.) It is perhaps a somewhat disappointing fact that E need not be the topological direct sum of M and N when they are provided with the induced topologies. Moreover, the natural algebraic isomorphism between E/M and N need not be a topological one.

PROPOSITION 29. *Let the convex space E be the algebraic direct sum of vector subspaces M and N, let p and q be the projection mappings of E onto M and N, and let h and k be the canonical mappings of E onto E/M and E/N. Then the following are equivalent:*

(i) *E is the topological direct sum of M and N;*

(ii) *p is continuous;*

(iii) *q is continuous;*

(iv) *h is an isomorphism of N onto E/M;*

(v) *k is an isomorphism of M onto E/N.*

Proof. Since $p + q$ is the identity mapping of E onto itself, (ii) and (iii) are equivalent. Now the direct sum topology on $M + N$ is the finest coinciding with the given topology on M and N; by Proposition 21 it coincides with the product topology and so is the coarsest making both p and q continuous. Hence (i) is equivalent to (ii) and (iii). Finally, for any neighbourhood U in E, $k(U \cap M)$ is a neighbourhood in E/N if and only if

$$(U \cap M) + N = p^{-1}(U \cap M)$$

is a neighbourhood in E. Thus (ii) is equivalent to (v) and similarly (iii) to (iv).

COROLLARY. *If the convex space E is the algebraic direct sum of the finite dimensional vector subspace M and the closed vector subspace N, then E is the topological direct sum of M and N.*

Proof. By Proposition 1, E/N is separated. Since k is a $(1, 1)$ continuous linear mapping of M onto E/N, M is also separated and so k is a topological isomorphism (Ch. II, Prop. 11), which proves (v).

We shall prove later (Ch. VI, Th. 8, Cor. 2) that if a Fréchet space E is the algebraic direct sum of the closed vector subspaces M and N then E is also their topological direct sum.

If E is a vector space and M a vector subspace, any vector subspace N of E such that E is the algebraic direct sum of M and N is called an *algebraic supplement* of M in E. It is always possible to find an algebraic supplement, for we have only to take a base A of M and extend it to form a base B of E; then $B \cap (\sim A)$ spans a vector subspace which is an algebraic supplement of M in E. Similarly, if E is a convex space and M a vector subspace, any vector subspace N of E such that E is the topological direct sum of M and N is called a *topological supplement* of M in E. Unfortunately, it is by no means always possible to find such a topological supplement. In the first place, if E is separated, M must be closed for a topological supplement to exist (for, by Proposition 22, M is closed in $M+N$). But not even all closed vector subspaces have topological supplements. Any two topological supplements of M in E are isomorphic, for both are isomorphic to E/M (Prop. 29).

PROPOSITION 30. *The vector subspace M of the convex space E has a topological supplement if and only if there is a continuous linear mapping p of E onto M such that $p^2 = p$.*

Proof. If N is a topological supplement of M in E and p is the projection mapping of $M+N$ onto M, then p has the required properties. Conversely, given such a mapping p, let $N = p^{-1}(o)$; then p is the projection mapping of $M+N$ onto M and (Prop. 29) E is the topological direct sum of M and N.

A linear mapping p of a vector space E into itself with the property $p^2 = p$ is called a *projector*; E is then the algebraic direct sum of $p(E)$ and $p^{-1}(o)$.

It follows from the Corollary to Proposition 29 that if M is a closed vector subspace of a convex space E for which E/M is finite-dimensional, then every algebraic supplement of M is also a topological supplement. Further:

PROPOSITION 31. *In a separated convex space, every finite-dimensional vector subspace has a topological supplement.*

Proof. Let $e_1, e_2, \ldots, e_n$ be a base for the finite-dimensional vector subspace M. There are continuous linear forms $f_1, f_2, \ldots, f_n$ such that $f_i(e_j) = 0$ when $i \neq j$ and $f_i(e_i) = 1$ (Ch. II, Lemma 5, Corollary, regarding the e_i as linear forms on the dual space). Write

$$p(x) = \sum_{1 \leqslant i \leqslant n} f_i(x)\, e_i.$$

Then

$$p^2(x) = \sum_{1 \leqslant j \leqslant n} f_j\Big(\sum_{1 \leqslant i \leqslant n} f_i(x)\, e_i\Big)\, e_j = \sum_{1 \leqslant i \leqslant n} \sum_{1 \leqslant j \leqslant n} f_i(x) f_j(e_i)\, e_j = p(x).$$

Hence p is a continuous projector onto M and so (Prop. 30) M has a topological supplement.

SUPPLEMENT

(1) *Finest convex topology.* If Γ is any set of indices, the set $\mathop{\times}\limits_{\gamma \in \Gamma} E_\gamma$, in which each E_γ is identical with a fixed set E, is usually denoted by E^Γ; it is the set of all functions defined on Γ and taking values in E. When E is a vector space, so is the product E^Γ. For example, the space of all scalar-valued functions on the set S (Ch. I, Suppl. 7) is the product Φ^S; moreover the topology of pointwise convergence on S is easily seen to be the product topology on Φ^S. We have already found it useful to regard the algebraic dual E^* of a vector space E as a vector subspace of Φ^E (Ch. III, Suppl. 3). Alternatively, if A is any base of E, E^* can be identified with Φ^A, and the weak topology $\sigma(E^*, E)$ is the product topology on Φ^A. The vector space E itself can be identified with a direct sum of copies of Φ, one for each member of the base A. The direct sum topology, the finest inducing the usual topology on each copy of Φ, is therefore the finest convex topology

$\tau(E, E^*)$. Another useful way of looking at $\tau(E, E^*)$ is to regard it as the inductive limit topology defined by all the finite-dimensional vector subspaces of E under their (unique) Euclidean topologies.

(2) *Associated separated topology.* As mentioned in Chapter I (see the remarks following Prop. 8) it is usually possible to avoid the study of non-separated convex spaces, by taking quotients. If E is not separated and N is the intersection of the neighbourhooods (the closure of the set consisting of the origin alone), then N is a closed vector subspace of E and so E/N is a separated convex space (Prop. 1). The canonical mapping k of E onto E/N, as well as mapping open sets onto open sets and compact sets onto compact sets, has the additional property that the image of a closed set is closed. (For if A is closed in E and $k(a) \notin k(A)$, there is an absolutely convex neighbourhood U with $(a+2U) \cap A = \emptyset$; since $N \subseteq U$, $(a+U+N) \cap A = \emptyset$ and it follows that $k(a)+k(U)$ does not meet $k(A)$.) A similar argument shows that E is complete if and only if E/N is complete. Any continuous linear mapping t of E into a separated convex space F must vanish on N (since $t(N)$ is contained in every neighbourhood in F) and so t has a canonical decomposition $t = u \circ k$, where u is a continuous linear mapping of E/N into F. Thus, in particular, E and E/N have the same dual.

(3) *Inductive limits.* Suppose that E is a vector space and that $(E_\gamma: \gamma \in \Gamma)$ is a family of vector subspaces, each E_γ being a convex space under the topology ξ_γ. Denote the inductive limit topology on E by ξ. If now $(F_\delta: \delta \in \Delta)$ is another family of vector subspaces of E, each F_δ having a convex topology η_δ, E can also be given the inductive limit topology η defined by these F_δ and it is often important to know the relationship between the topologies ξ and η. One simple criterion is the following: if, to each E_γ corresponds an F_δ such that $E_\gamma \subseteq F_\delta$ and η_δ is coarser than ξ_γ on E_γ, then η is coarser than ξ. (For this condition ensures that η induces on each E_γ a topology coarser than ξ_γ.) If the symmetrically opposite condition with (E_γ) and (F_δ) interchanged is also satisfied (so that the spaces (E_γ) and (F_δ) together with their topologies interlace) then ξ and η are identical.

A useful consequence of this is that the inductive limit of the family $(E_\gamma: \gamma \in \Gamma)$ of subspaces is the same as the inductive limit of the subfamily $(E_\gamma: \gamma \in \Delta)$, where $\Delta \subseteq \Gamma$, provided that each E_γ is contained in some E_δ with $\delta \in \Delta$, and that the topology on E_γ induced by that of E_δ is coarser than the original topology on E_γ. Thus, for example, in defining the topology on $\mathscr{K}(S)$, where S is locally compact (Ch. I, Suppl. 2*c*) we can take any base $\mathscr{A}$ of compact subsets of S (with the property that every compact set in S is contained in one of the sets of $\mathscr{A}$), and then take the inductive limit of the subspaces $\mathscr{K}_A(S)$ with $A \in \mathscr{A}$. As another illustration of this property, we point out that the topology of $\mathscr{D}$ (Ch. I, Suppl. 3*d*) is independent of the particular sequence of intervals used to define it. (Previously we used the intervals $[-n, n]$, but any base of compact intervals would give the same topology.)

(4) *Inductive and projective limits of extreme topologies.* We consider here how the topologies σ, τ and β behave under inductive and projective limit operations. There are only two completely general results: any (separated) inductive limit of spaces with the topology τ also has the topology τ and any projective limit of spaces with the topology σ also has the topology σ. (If E is the inductive limit of the spaces E_γ by the mappings u_γ, the inductive limit topology ξ is the finest making the u_γ continuous. But if E' is the dual of E and each E_γ has the topology $\tau(E_\gamma, E'_\gamma)$, the finer topology $\tau(E, E')$ also makes the u_γ continuous (Ch. III, Prop. 14) and so ξ is identical with $\tau(E, E')$. The proof of the second assertion runs similarly, using Ch. II, Prop. 13.)

Thus a weak topology induces a weak topology on a vector subspace and a product of weak topologies is a weak topology. Also a separated quotient of a weak topology is a weak topology (by Prop. 3 and direct examination of the neighbourhoods). But a direct sum of weak topologies may fail to be a weak topology ($\tau(E, E^*)$ provides a counter-example; see (1)).

The other general result shows that the topology τ is preserved under taking separated quotients or direct sums. The topology τ, and also the topology β, are preserved under products. (If $E = \Sigma E_\gamma$ and $E' = \mathsf{X} E'_\gamma$, then any $\sigma(E, E')$-bounded set is contained in a finite sum or union of sets bounded in individual E_γ, by Proposition 24; the corollary shows that the same is true of absolutely convex $\sigma(E, E')$-compact sets. Hence, by

CHAPTER VI

COMPLETENESS AND THE CLOSED GRAPH THEOREM

In this and in the succeeding chapters, each section contains its own introductory remarks.

It is the idea of completeness, of one kind or another, that forms the unifying feature of the subject matter of this chapter.

1. The completion of a convex space. In this section we show how any separated convex space E can be embedded as a dense vector subspace in a complete separated convex space $\hat{E}$, called the *completion* of E. Every Cauchy filter in E, being the base of a Cauchy filter in $\hat{E}$, is then convergent to a point of $\hat{E}$. There are various ways of constructing such a completion $\hat{E}$, of which we give one, and so, strictly speaking, $\hat{E}$ ought to be described only as *a* completion of E. But it turns out that all methods give the same completion (up to isomorphisms; this is proved after Proposition 6, Corollary 1) and so from the outset we shall call $\hat{E}$ *the* completion of E. Theorem 3 below gives the main result for the completion of E. Since the topology of any separated convex space may be regarded as the topology of uniform convergence on the equicontinuous subsets of its dual, it is slightly more general to consider the completion of a dual space under a topology of $\mathscr{A}$-convergence. We suppose, for technical simplicity and without loss of generality, that the polars of the sets of $\mathscr{A}$ form a base of neighbourhoods in the topology of $\mathscr{A}$-convergence and that this topology is separated. This is ensured by the conditions B 1–B 3 of Chapter III, § 2.

THEOREM 1. *Let* (E, E') *be a dual pair and let* $\mathscr{A}$ *be a set of bounded subsets of* E *satisfying* B 1–B 3 (Ch. III, § 2). *Then the completion of* E' *under the topology of* $\mathscr{A}$*-convergence is the vector subspace*

$$M' = \bigcap_{A \in \mathscr{A}} (E' + A^0)$$

of E^*, under the topology of $\mathscr{A}$-convergence, the polars A^0 being taken in E^*.

Proof. If $z' \in M'$ and $A \in \mathscr{A}$, $z' \in E' + A^0$ and so there is some $x' \in E'$ with $z' \in x' + A^0$. Hence each $A \in \mathscr{A}$ is $\sigma(E, M')$-bounded and the topology of $\mathscr{A}$-convergence can be defined on M'. Also E' is dense in M' under this topology, for every neighbourhood $z' + A^0 = z' - A^0$ of z' contains a point x' of E'.

The theorem will therefore be proved if we show that M' is complete. Let $\mathscr{F}'$ be a Cauchy filter in M'. Since the topology of $\mathscr{A}$-convergence is finer than the topology $\sigma(M', E)$ induced on M' by $\sigma(E^*, E)$, $\mathscr{F}'$ is the base of a $\sigma(E^*, E)$-Cauchy filter in E^*. But E^* is $\sigma(E^*, E)$-complete (Ch. III, Prop. 13) and so there is a point $x^* \in E^*$ with $\mathscr{F}' \to x^*$ in $\sigma(E^*, E)$. Now for each $A \in \mathscr{A}$, $\mathscr{F}'$ contains a set B' small of order A^0; if $z' \in B'$, $z' + A^0$ is a $\sigma(E^*, E)$-closed set belonging to $\mathscr{F}'$. Hence

$$x^* \in z' + A^0 \subseteq E' + A^0 + A^0 = E' + 2A^0$$

and so $x^* \in M'$. Also from $z' \in x^* + A^0$ it follows that $B' \subseteq x^* + A^0$ and therefore $\mathscr{F}' \to x^*$ in the topology of $\mathscr{A}$-convergence. Thus M' is complete.

COROLLARY 1. *Each $A \in \mathscr{A}$ is $\sigma(E, M')$-bounded.*

COROLLARY 2. *If E' has the topology $\beta(E', E)$, its completion M has the topology $\beta(M', E)$.*

There are two other useful characterisations of the completion of E' under the topology of $\mathscr{A}$-convergence. For them, it is convenient to assume that each $A \in \mathscr{A}$ is $\sigma(E, E')$-closed and absolutely convex. This we may always do, because the topology of $\mathscr{A}$-convergence is the same as the topology of uniform convergence on the set of closed absolutely convex envelopes of the sets of $\mathscr{A}$. We require first:

LEMMA 1. *Let A be an absolutely convex subset of the convex space E and let t be any linear mapping of E into another convex space F. Then t is continuous on A under the induced topology if (and only if) t is continuous on A at the origin.*

Proof. Let $a \in A$ and let V be any neighbourhood in F. If t is continuous on A at the origin, there is an absolutely convex neighbourhood U in E with $t(U \cap A) \subseteq \frac{1}{2}V$. Then if

$$x \in (a+U) \cap A, \quad x-a \in (2U) \cap (A-A) \subseteq 2(U \cap A)$$

and so $t(x) \in t(a)+V$. Thus t is continuous at a.

THEOREM 2. *Let ξ be any topology of the dual pair (E, E') and let $\mathscr{A}$ be a set of closed absolutely convex bounded subsets of E satisfying* B 1–B 3. *Then the completion M' of E' under the topology of $\mathscr{A}$-convergence is the set of all linear forms on E which are ξ-continuous on each $A \in \mathscr{A}$.*

Proof. Let $\mathscr{U}$ be a base of closed absolutely convex ξ-neighbourhoods; then

$$E' = \bigcup_{U \in \mathscr{U}} U^0$$

and so z' belongs to

$$M' = \bigcap_{A \in \mathscr{A}} (E' + A^0)$$

if and only if for each $A \in \mathscr{A}$ there is some $U \in \mathscr{U}$ with

$$z' \in U^0 + A^0.$$

Next we show that $(U \cap A)^0 \subseteq U^0 + A^0 \subseteq 2(U \cap A)^0$. Here the second inclusion is a consequence of $U^0 \subseteq (U \cap A)^0$ and $A^0 \subseteq (U \cap A)^0$. For the first, U^0 is $\sigma(E^*, E)$-compact (Ch. III, Th. 6) and A^0 is $\sigma(E^*, E)$-closed; hence (Ch. III, Lemma 7 (iii)) $U^0 + A^0$ is $\sigma(E^*, E)$-closed. Thus $U^0 + A^0$ contains the $\sigma(E^*, E)$-closed absolutely convex envelope of $U^0 \cup A^0$, which is the polar of $U \cap A$ (Ch. II, Th. 4, Cor. 3).

Now if z' is continuous on the $A \in \mathscr{A}$, then for each A there is some $U \in \mathscr{U}$ with $z' \in (U \cap A)^0$ and so $z' \in M'$. Conversely, let $z' \in M'$ and $A \in \mathscr{A}$. Then for each $\epsilon > 0$, $(2/\epsilon) A \in \mathscr{A}$ (by B 2) and so there is some $U \in \mathscr{U}$ with

$$z' \in ((2/\epsilon)\, U)^0 + ((2/\epsilon)\, A)^0 = \tfrac{1}{2}\epsilon(U^0 + A^0) \subseteq \epsilon(U \cap A)^0.$$

Hence z' is continuous at the origin on A and therefore continuous on A by Lemma 1.

COROLLARY 1. *If every linear form on E continuous on each $A \in \mathscr{A}$ is also continuous on E, then E' is complete under the topology of $\mathscr{A}$-convergence.*

Corollary 2. *The same linear forms are continuous on each* $A \in \mathscr{A}$ *for all topologies of the dual pair* (E, E').

For M' does not depend on ξ.

Corollary 3. *The topologies* $\sigma(E, E')$ *and* $\sigma(E, M')$ *coincide on each* $A \in \mathscr{A}$.

Proof. Take $\sigma(E, E')$ for ξ in Theorem 2. The ξ-continuity of each $z' \in M'$ at a point $a \in A$ ensures that for each $\sigma(E, M')$-neighbourhood V, there is a $\sigma(E, E')$-neighbourhood U with

$$(a+U) \cap A \subseteq a+V.$$

Since $\sigma(E, M')$ is finer than $\sigma(E, E')$, they coincide on each $A \in \mathscr{A}$.

Corollary 4. *If* E' *has the topology* $\tau(E', E)$, *its completion* M' *has the topology* $\tau(M', E)$.

For if $\mathscr{A}$ is the set of all absolutely convex $\sigma(E, E')$-compact sets of E, by Corollary 3 $\mathscr{A}$ is also the set of all absolutely convex $\sigma(E, M')$-compact sets of E.

Proposition 1. *Suppose that* E *is a separated Mackey space, and that every compact subset of* E *is contained in some* $A \in \mathscr{A}$. *Then the dual* E' *of* E *is complete under the topology of* $\mathscr{A}$*-convergence.*

Proof. Let z' be continuous on each $A \in \mathscr{A}$; we prove that z' is a bounded linear form on E (cf. Ch. v, § 3). If not, there are points x_n belonging to a bounded subset B of E with $|\langle x_n, z' \rangle| > n^2$. Since B is bounded, $n^{-1}x_n \to o$ and so the set consisting of the points $n^{-1}x_n$ and the origin is compact. There is therefore some $A \in \mathscr{A}$ with $n^{-1}x_n \in A$ and so z' is unbounded on A, contrary to Theorem 1, Corollary 1.

Corollary 1. *The dual* E' *of a separated Mackey space* E *is complete under* $\beta(E', E)$. *In particular, the dual* E' *of a metrisable space* E *is complete under* $\beta(E', E)$.

Corollary 2. *The dual* E' *of a complete separated Mackey space* E *is complete under* $\tau(E', E)$.

For every compact subset B of E is then contained in an absolutely convex $\sigma(E, E')$-compact set (its closed absolutely convex envelope: Ch. III, Th. 5, Corollary).

We now give the second characterisation of the completion of a dual space.

Proposition 2. *Let (E, E') be a dual pair and let $\mathscr{A}$ be a set of weakly closed absolutely convex bounded subsets of E satisfying* B 1–B 3. *Let M' be the completion of E' under the topology of $\mathscr{A}$-convergence. Then $z' \in M'$ if and only if $(z'^{-1}(0)) \cap A$ is $\sigma(E, E')$-closed for each $A \in \mathscr{A}$.*

Proof. If $z' \in M'$, z' is continuous on each $A \in \mathscr{A}$, by Theorem 2, and so $z'^{-1}(0) \cap A$ is closed in A. Since A is closed, this set is also closed in E.

Conversely, let $H = z'^{-1}(0)$ intersect each $A \in \mathscr{A}$ in a closed set and let $\epsilon > 0$ be given. If z' does not vanish on A there is a point $a \in A$ with $\langle a, z' \rangle = \alpha$, where $0 < \alpha \leqslant \epsilon$. Now $a \notin H$ and so there is some $\sigma(E, E')$-neighbourhood U with $a + U$ not meeting the closed set $H \cap (2A)$. Then $|\langle x, z' \rangle| < \alpha \leqslant \epsilon$ on $U \cap A$, for otherwise there would be some $x \in U \cap A$ with $\langle x, z' \rangle = -\alpha$ and $x + a$ would belong to $H \cap (2A) \cap (a + U)$. Thus z' is continuous on each A and so belongs to M'.

Since the completion of E' under the topology of $\mathscr{A}$-convergence is a vector subspace M' of E^*, it is possible to consider inclusion relations between completions of E' under different topologies. Theorem 1 gives immediately:

Proposition 3. *Let (E, E') be a dual pair and let $\mathscr{A}$ and $\mathscr{B}$ be two sets of bounded subsets of E each satisfying* B 1–B 3. *Let M' and N' be the completions of E' under the topologies of $\mathscr{A}$- and $\mathscr{B}$-convergence respectively. Then if $\mathscr{A} \subseteq \mathscr{B}$, $N' \subseteq M'$. In particular, if E' is complete under the topology of $\mathscr{A}$-convergence, it is also complete under the (finer) topology of $\mathscr{B}$-convergence.*

Corollary. *If also K' is a subset of E' which is complete under the topology of $\mathscr{A}$-convergence, K' is complete under the topology of $\mathscr{B}$-convergence.*

Proof. Since K' is closed in M', it is closed in N' under the topology induced on it by that of M'. But this topology is coarser than the topology of $\mathscr{B}$-convergence on N' and so K' is closed in N'. Hence K' is complete for the topology of $\mathscr{B}$-convergence.

We can use this Corollary to sharpen the result (Ch. III, Th. 6) that the polar of a neighbourhood is weakly compact. If ξ is any topology of the dual pair (E, E') let us denote by $\pi(\xi)$ the topology on E' of uniform convergence on the set of all closed absolutely convex precompact subsets of E.

PROPOSITION 4. *Suppose that E is a separated convex space with topology ξ and dual E'. Then for every neighbourhood U in E, U^0 is $\pi(\xi)$-compact in E'; also $\pi(\xi)$ is the finest polar topology under which the sets U^0 are compact (or even precompact).*

Proof. We already know (Ch. III, Th. 6) that U^0 is $\sigma(E', E)$-compact and therefore $\sigma(E', E)$-complete. By the Corollary to Proposition 3, U^0 is also $\pi(\xi)$-complete. Also (Ch. III, Th. 3, Cor. 1) U^0 is $\pi(\xi)$-precompact and so is $\pi(\xi)$-compact. Finally, if each U^0 is precompact for the topology of $\mathscr{A}$-convergence, each $A \in \mathscr{A}$ must be precompact under ξ (Ch. III, Th. 3, Cor. 1).

So far in this section we have been considering the completion of a dual space. We now specialise these results to a convex space with a given topology.

THEOREM 3. *Suppose that E is a separated convex space with dual E', and let $\mathscr{U}$ be a base of neighbourhoods. Then the completion $\hat{E}$ of E is the subspace $\bigcap_{U \in \mathscr{U}} (E + U^{00})$ of E'^*, the bipolars being taken in E'^*, and the sets $U^{00} \cap \hat{E} (U \in \mathscr{U})$, form a base of neighbourhoods in $\hat{E}$. The completion $\hat{E}$ of E is the set of all linear forms z on E' which are $\sigma(E', E)$-continuous on the polar of each $U \in \mathscr{U}$, or, equivalently, for which $(z^{-1}(0)) \cap U^0$ is $\sigma(E', E)$-closed for each $U \in \mathscr{U}$.*

Proof. These assertions follow from Theorems 1 and 2 and Proposition 2; since $\hat{E}$ has the topology of uniform convergence on the sets U^0 $(U \in \mathscr{U})$, the sets $U^{00} \cap \hat{E}$ form a base of neighbourhooods.

In the statement of the theorem, $\sigma(E', E)$ may be replaced by any topology of the dual pair (E', E) (Th. 2, Cor. 2 and Ch. II, Prop. 8). Theorem 2, Corollaries 3 and 4, show that the topologies $\sigma(E', E)$ and $\sigma(E', \hat{E})$ coincide on each U^0 and that if E has the topology $\tau(E, E')$, $\hat{E}$ has the topology $\tau(\hat{E}, E')$. Also

Corollary 1. *The space E is complete if and only if every linear form on E' that is $\sigma(E',E)$-continuous on each U^0 is $\sigma(E',E)$-continuous on E'.*

Corollary 2. *The space E is complete if and only if every hyperplane in E' that intersects each U^0 in a $\sigma(E',E)$-closed set is $\sigma(E',E)$-closed.*

From Proposition 3 we deduce at once:

Proposition 5. *If E is a complete separated convex space, then E is complete under any finer topology of the same dual pair.*

Proposition 6. *Let E and F be separated convex spaces and let t be a continuous linear mapping of E into F. Then t has a unique extension $\hat{t}$ which is a continuous linear mapping of $\hat{E}$ into $\hat{F}$.*

Proof. Let t' be the transpose of t, mapping F' into E', and let t'^* be the transpose of t', mapping E'^* into F'^*. Then, if $\mathscr{U}$ and $\mathscr{V}$ are bases of closed absolutely convex neighbourhoods in E and F, to each $V \in \mathscr{V}$ corresponds some $U \in \mathscr{U}$ with $t(U) \subseteq V$. Taking bipolars in E'^* and F'^*, we have $t'^*(U^{00}) \subseteq V^{00}$. Hence

$$t'^*(\hat{E}) \subseteq \bigcap_{U \in \mathscr{U}} t'^*(E+U^{00}) \subseteq \bigcap_{V \in \mathscr{V}} (F+V^{00}) = \hat{F},$$

and $t'^*(\hat{E} \cap U^{00}) \subseteq \hat{F} \cap V^{00}$. Thus the restriction $\hat{t}$ of t'^* to $\hat{E}$ is a continuous linear mapping of $\hat{E}$ into $\hat{F}$ extending t. To show that $\hat{t}$ is unique, suppose that u is another continuous linear extension of t to a mapping of $\hat{E}$ into $\hat{F}$. Then $v = \hat{t} - u$ is continuous on $\hat{E}$ and $v(E) = \{o\}$. Hence $v(\hat{E}) = \overline{\{o\}} = \{o\}$, and so $\hat{t} = u$ on $\hat{E}$.

Corollary 1. *If also t is an isomorphism of E onto $t(E)$, then $\hat{t}$ is an isomorphism of $\hat{E}$ onto $\hat{t}(\hat{E})$.*

Proof. Since $\hat{t}$ is continuous, $\hat{t}(\hat{E})$ is contained in the closure (in $\hat{F}$) of $t(E)$, and since t is an isomorphism, the sets $\overline{t(U)}$, as U runs through $\mathscr{U}$, form a base of neighbourhoods for $\hat{t}(\hat{E})$. It will be sufficient to prove that if $x \in \hat{E}$ and $\hat{t}(x) \in \overline{t(U)}$ then $x \in 2\overline{U}$. For then $\hat{t}(x) = o$ implies that $x \in 2\overline{U}$ for all $U \in \mathscr{U}$ and therefore $x = o$. Thus $\hat{t}$ is $(1,1)$; the relation $\hat{t}^{-1}(\overline{t(U)}) \subseteq 2\overline{U}$ shows that, on $\hat{t}(\hat{E})$, $\hat{t}^{-1}$ is continuous.

So suppose that for some $x \in \hat{E}$, $\hat{t}(x) \in \overline{t(U)}$ but $x \notin 2\overline{U}$. Then there is some neighbourhood of x, which we may suppose contained in $x+\overline{U}$, which does not meet $2U$, and so E contains a point y in $x+\overline{U}$ but not in $2U$. Then

$$t(y) \in \hat{t}(x)+\hat{t}(\overline{U}) \subseteq \overline{t(U)}+\overline{t(U)} = 2\overline{t(U)}.$$

But $t(U)$ is closed in $t(E)$ (U being closed and t an isomorphism) and so $t(y) \in 2t(U)$, which contradicts $y \notin 2U$.

This corollary enables us to justify the assertion, made at the beginning of the section, that completion is unique up to an isomorphism. For suppose that E can be mapped, by an isomorphism t say, onto a dense vector subspace of a complete separated convex space F. Then $\hat{E}$ is isomorphic to $\hat{t}(\hat{E})$, which is therefore complete, and so closed in F. Since it contains the dense vector subspace $t(E)$, it must be identical with F and therefore $\hat{E}$ and F are isomorphic.

Corollary 2. *If E is a separated convex space with dual E', then the dual of $\hat{E}$ is also E'.*

For every continuous linear form on E has a unique continuous extension to $\hat{E}$, the scalar field being complete, and this defines an isomorphism between the duals of E and $\hat{E}$.

It follows that if U is a closed absolutely convex neighbourhood in the separated convex space E then $U^{00} \cap \hat{E}$ is the closure of U in $\hat{E}$ (Ch. II, Th. 4 and Prop. 8). If p is any continuous seminorm on E, the gauge of the closed absolutely convex neighbourhood U, say, then the gauge $\hat{p}$ of $U^{00} \cap \hat{E}$ is given by

$$\hat{p}(z) = \sup\{|\langle z, x'\rangle| : x' \in U^0\}$$

and so $\hat{p}$ is the (unique) extension by continuity of p to $\hat{E}$.

Thus any normed space can be embedded in its completion, which is a Banach space under the extended norm. (Cf. Ch. V, Props. 16 *et seq.*)

It is clear that the completion of a metrisable convex space is also metrisable, and therefore a Fréchet space.

Proposition 7. *The completion of a separated barrelled space is barrelled.*

Proof. Let B be a barrel in the completion $\hat{E}$ of the barrelled space E. Then $B \cap E$ is a barrel, and therefore a neighbourhood in E, and so its closure B in $\hat{E}$ is a neighbourhood. Thus $\hat{E}$ is barrelled.

We end this section by applying Theorem 3 to prove a result on weak compactness. The value of this result lies in the fact that it enables the weak compactness of a set to be deduced from purely sequential properties. Given a sequence (x_n) of points of a topological space, we shall say that the point a is a *cluster point* of the sequence if each neighbourhood of a contains points x_n with arbitrarily large n. In a metric space this clearly implies the existence of a subsequence convergent to a, and it follows that in a metric space the set A has a compact closure if and only if every sequence of points of A has a cluster point. In a general topological space this last result fails, though a subset A of a convex space from which every sequence has a cluster point is always precompact. (If not, there would be an absolutely convex neighbourhood U and a sequence (x_n) of points of A with $x_m - x_n \notin U$ for $n \neq m$ and so (x_n) could not have a cluster point.) Thus in a complete convex space a set A has a compact closure provided that every sequence of points from it has a cluster point. Much more than this is true however:

THEOREM 4. (Eberlein's theorem.) *Let E be a complete separated convex space. If every sequence of points of the subset A of E has a weak cluster point, then the weak closure of A is weakly compact.*

Proof. Let E' be the dual of E. The closure B of A in E'^* under $\sigma(E'^*, E')$ is compact, A being precompact and E'^* complete under this topology; let $z \in B$. To prove that $z \in E$, it is sufficient, since E is complete, to show that z is continuous at the origin on the polar of each neighbourhood U (Th. 3 and Lemma 1). Suppose not. Then for some $\epsilon > 0$, there is no $\sigma(E', E)$-neighbourhood V' with $z \in \epsilon(V' \cap U^0)^0$. Thus there is some $x'_1 \in U^0$ with $|\langle z, x'_1\rangle| > \epsilon$. Since $z \in B$, there is some $x_1 \in A$ with

$$|\langle z - x_1, x'_1\rangle| < \tfrac{1}{3}\epsilon;$$

then there is some $x'_2 \in U^0$ with

$$|\langle x_1, x'_2\rangle| \leqslant \tfrac{1}{3}\epsilon \quad \text{and} \quad |\langle z, x'_2\rangle| > \epsilon,$$

and so on. We thus define sequences (x_n) in A and (x'_n) in U^0 with $|\langle z - x_n, x'_i\rangle| < \frac{1}{3}\epsilon$ for $i \leqslant n$, $|\langle x_i, x'_n\rangle| \leqslant \frac{1}{3}\epsilon$ for $i < n$ and $|\langle z, x'_n\rangle| > \epsilon$.

Now under $\sigma(E,E')$, (x_n) has a cluster point a, and since U^0 is $\sigma(E',E)$-compact, (x'_n) has a cluster point a'. Then for each i,

$$\langle a, x'_i \rangle = \lim_{j\to\infty} \langle x_{n(j)}, x'_i \rangle$$

for some sequence $(n(j))$ and so $|\langle z, x'_i\rangle - \langle a, x'_i\rangle| \leqslant \frac{1}{3}\epsilon$. Thus $|\langle a, x'_i\rangle| > \epsilon - \frac{1}{3}\epsilon = \frac{2}{3}\epsilon$ for all i, and therefore

$$|\langle a, a'\rangle| = \lim_{r\to\infty} |\langle a, x'_{i(r)}\rangle| \geqslant \tfrac{2}{3}\epsilon.$$

But also for each i,

$$\langle x_i, a'\rangle = \lim_{j\to\infty} \langle x_i, x'_{n(j)}\rangle$$

for some sequence $(n(j))$ and so $|\langle x_i, a'\rangle| \leqslant \frac{1}{3}\epsilon$, whence

$$|\langle a, a'\rangle| = \lim_{r\to\infty} |\langle x_{i(r)}, a'\rangle| \leqslant \tfrac{1}{3}\epsilon.$$

This contradiction proves that z is in E; hence, under $\sigma(E,E')$, the closure of A is B and thus is compact.

Corollary. *Let E be a separated convex space with dual E' and let A be a subset of E with the property that every sequence of points of A has a cluster point. If E is complete under $\tau(E,E')$, then, under the given topology, $\bar{A}$ is compact.*

Proof. Every sequence of points of A certainly has a $\sigma(E,E')$-cluster point, and so, by the theorem, its $\sigma(E,E')$-closure, B say, is $\sigma(E,E')$-compact. Therefore B is $\sigma(E,E')$-complete and so complete in the given topology (Prop. 3, Corollary). Hence the closed subset $\bar{A}$ of B is complete. But A, and so also $\bar{A}$, is precompact; thus $\bar{A}$ is compact.

2. Closed graph and open mapping theorems. If E and F are convex spaces and t is a continuous (1, 1) linear mapping of E onto F, it is natural to ask what additional conditions ensure that the inverse mapping t^{-1} is also continuous, that is, that t is an isomorphism of E onto F. It follows from a theorem of Banach that t is an isomorphism whenever E and F are both Fréchet spaces. Another closely related theorem of Banach proves that if a linear mapping t of one Fréchet space into another has a closed graph, then t is continuous. These theorems can be extended to wider classes of convex spaces and their generalisations form the subject matter of this section.

It turns out that there is still a third theorem of Banach that impinges on these ideas. In the previous section we proved that a hyperplane in the dual of a complete separated convex space is weakly closed if it has weakly closed intersections with the polars of the neighbourhoods. Banach's theorem shows that, for a Banach space, this property holds not only for hyperplanes, but for all vector subspaces of its dual (cf. Th. 5 below and Suppl. 2). We approach the closed graph and open mapping theorems by way of the class of convex spaces with this property.

Let E be a separated convex space with dual E'. We shall call the subset A' of E' *nearly closed* if $A' \cap U^0$ is $\sigma(E', E)$-closed for every neighbourhood U in E. Then E is called *fully complete* if every nearly closed vector subspace of E' is $\sigma(E', E)$-closed. The name is justified by Theorem 3, Corollary 2, which shows that every fully complete space is complete. Fully complete spaces share with complete spaces the following property, which is immediate from the definition:

Proposition 8. *A fully complete space remains fully complete under any finer topology of the same dual pair.*

There are complete separated convex spaces that are not fully complete (see Suppl. 1), but this cannot happen for metrisable spaces. To prove this, we use:

Lemma 2. *Suppose that U is a closed absolutely convex neighbourhood in the separated convex space E, and that A' is a $\sigma(E', E)$-closed absolutely convex subset of its dual E'. Then*

$$(A' \cap U^0)^0 \subseteq A'^0 + 2U,$$

the polars being taken in E.

Proof. The set $(A' \cap U^0)^0$ is the closed absolutely convex envelope of $A'^0 \cup U^{00} = A'^0 \cup U$ and this is a subset of $\overline{A'^0 + U}$, which in turn is contained in $A'^0 + 2U$.

Theorem 5. *Every Fréchet space is fully complete.*

Proof. Let E be a Fréchet space and M' a nearly closed vector subspace of its dual E'. It is sufficient to prove that, for each closed absolutely convex neighbourhood U, there is a neighbourhood V with $(M' \cap V^0)^0 \subseteq M'^0 + U$. For then

$$(M'^0 + U)^0 \subseteq (M' \cap V^0)^{00} = M' \cap V^0,$$

since M' is nearly closed. But since M' is a vector subspace,

$$(M'^0 + U)^0 = (M'^0 \cup U)^0 = M'^{00} \cap U^0$$

and so $M'^{00} \cap U^0 \subseteq M'$. Since each point of E' is in some U^0, this implies that $M'^{00} \subseteq M'$, and therefore M' is $\sigma(E', E)$-closed.

To prove the existence of V, take a base (U_n) of closed absolutely convex neighbourhoods with $U_1 = U$ and $U_{n+1} \subseteq \frac{1}{2}U_n$ for $n = 1, 2, \ldots$. Then for each n, $A'_n = M' \cap U_n^0$ is $\sigma(E', E)$-closed and $A'_n = A'_{n+1} \cap U_n^0$. Hence by Lemma 2, $A'^0_n \subseteq A'^0_{n+1} + 2U_n$. Thus if $a \in A'^0_1$, $a \in A'^0_2 + 2U_1$ and so there is a point $x_1 \in 2U_1$ with $a - x_1 \in A'^0_2$. Similarly there is a point $x_2 \in 2U_2$ with

$$a - x_1 - x_2 \in A'^0_3,$$

and, continuing thus, we obtain a sequence (x_n) with $x_n \in 2U_n$ and

$$a - \sum_{1 \leqslant r \leqslant n} x_r \in A'^0_{n+1}.$$

Now the series Σx_r is convergent, since E is complete and

$$\sum_{n \leqslant r \leqslant n+p} x_r \in 2U_n + 2U_{n+1} + \ldots + 2U_{n+p} \subseteq 4U_n;$$

let
$$x_0 = \sum_{n=1}^{\infty} x_n,$$

so that $x_0 \in 4U_1$. Also if $x' \in M'$, then $x' \in A'_n$ for all sufficiently large n and so

$$|\langle a - x_0, x' \rangle| = \lim_{n \to \infty} |\langle a - \sum_{1 \leqslant r \leqslant n} x_r, x' \rangle| \leqslant 1.$$

Hence

$$a \in x_0 + M'^0 \subseteq 4U_1 + M'^0 \quad \text{and so} \quad A'^0_1 = (M' \cap U^0)^0 \subseteq M'^0 + 4U.$$

Thus $V = \frac{1}{4}U$ will suffice and the theorem is proved.

If E and F are topological spaces, the mapping f of E into F is called *open* if $f(A)$ is open for each open set A. Thus the inverse f^{-1} of a (1, 1) open mapping f of E onto F is continuous, and a continuous open (1, 1) mapping of E onto F is a homeomorphism.

It is also clear that if f is a continuous open mapping of E onto F, then the topology on F is the finest under which f is continuous.

When E and F are convex spaces and t is a linear mapping of E into F, then t is open if and only if $t(U)$ is a neighbourhood in F for each neighbourhood U in E (and so an open mapping into F is necessarily *onto* F: a continuous linear mapping t of E into F for which $t(A)$ is open in $t(E)$ for every open set A in E is sometimes called a *homomorphism*). For example, the canonical mapping of a convex space E onto a quotient E/M is open.

It often happens that, for a linear mapping t of E onto F, although we cannot assert that $t(U)$ is a neighbourhood, we can easily prove that $\overline{t(U)}$ is. This is the case when F is a barrelled space, for $\overline{t(U)}$ is a barrel for each absolutely convex neighbourhood U in E. It is convenient, for the rest of this section, to make the following definition. We shall call the linear mapping t of the convex space E into the convex space F *nearly open* if $\overline{t(U)}$ is a neighbourhood in F for every neighbourhood U in E.

Lemma 3. *Let E and F be separated convex spaces with duals E' and F'. If t is a nearly open linear mapping of E into F, with transpose t' (mapping F' into E^*), and N' is a nearly closed vector subspace of F', then $t'(N') \cap E'$ is nearly closed.*

Proof. If U is a neighbourhood in E, then

$$t'(N') \cap E' \cap U^0 = t'(N') \cap U^0 = t'(N' \cap t'^{-1}(U^0)).$$

Now $t'^{-1}(U^0) = (t(U))^0$ (Ch. II, Lemma 6 applied to the dual pairs (E, E^*) and (F, F')); since t is nearly open, $\overline{t(U)}$ is a neighbourhood in F and so $(t(U))^0 = (\overline{t(U)})^0$ is $\sigma(F', F)$-compact. Hence since N' is nearly closed, $N' \cap t'^{-1}(U^0)$ is $\sigma(F', F)$-compact. But t' is continuous under the topologies $\sigma(F', F)$ and $\sigma(E^*, E)$; hence $t'(N') \cap U^0$ is $\sigma(E^*, E)$-compact and so $\sigma(E', E)$-compact. Thus $t'(N') \cap E'$ is nearly closed.

Proposition 9. *If there is a continuous (nearly) open linear mapping of a fully complete space onto a separated convex space F, then F is fully complete.*

Proof. Let t be a continuous nearly open linear mapping of the fully complete space E onto F, and let N' be a nearly closed vector subspace of the dual F' of F. Since t is continuous, t' maps F' into the dual E' of E and, by Lemma 3, $t'(N')$ is nearly closed. Hence $t'(N')$ is weakly closed since E is fully complete. Now

$$N' = t'^{-1}(t'(N'))$$

since $t'^{-1}(o) = (t(E))^0 = \{o\}$, t mapping E onto F; thus N' is weakly closed since t' is weakly continuous. Thus F is fully complete.

COROLLARY. *The quotient of a fully complete space by a closed vector subspace is fully complete.*

The *graph* of any mapping f of one set E into another set F is the subset of $E \times F$ consisting of all elements of the form $(x, f(x))$ with $x \in E$. If E and F are topological spaces with F separated, and if f is continuous, then its graph G is closed. For if $(x, y) \notin G$ there are disjoint neighbourhoods U of $f(x)$ and V of y, and $(f^{-1}(U), V)$ is a neighbourhood of (x, y) not meeting G.

When E and F are convex spaces and t is a linear mapping of E into F, the graph of t is a vector subspace of $E \times F$. We shall use the following condition for the graph to be closed:

LEMMA 4. *Let E and F be separated convex spaces with duals E' and F' and let t be a linear mapping of E into F, with transpose t' (mapping F' into E^*). Then the graph of t is closed if and only if $t'^{-1}(E')$ is dense in F' under $\sigma(F', F)$.*

Proof. The set $t'^{-1}(E')$ is dense in F' under $\sigma(F', F)$ if and only if $(t'^{-1}(E'))^0 = \{o\}$. Now if $\mathscr{U}$ is a base of absolutely convex neighbourhoods in E, then

$$E' = \bigcup_{U \in \mathscr{U}} U^0$$

and so

$$t'^{-1}(E') = \bigcup_{U \in \mathscr{U}} t'^{-1}(U^0) = \bigcup_{U \in \mathscr{U}} (t(U))^0.$$

Hence

$$(t'^{-1}(E'))^0 = \bigcap_{U \in \mathscr{U}} (t(U))^{00} = \bigcap_{U \in \mathscr{U}} \overline{t(U)}.$$

But y is in this last set if and only if, for each $U \in \mathscr{U}$ and each neighbourhood V in F, $y + V$ meets $t(U)$, i.e. $(o, y) \in \overline{G}$, where G is the graph of t.

Hence if G is closed, $(t'^{-1}(E'))^0 = \{o\}$. Conversely, if

$$(t'^{-1}(E'))^0 = \{o\},$$

and if $(x,y) \in \bar{G}$, then $(o, y-t(x)) \in \bar{G}$ and so $y-t(x) = o$; thus $(x,y) \in G$ and G is closed.

COROLLARY. *If the graph of t is closed, then $t^{-1}(o)$ is closed.*

For, putting $N' = t'^{-1}(E')$, $N'^0 = \{o\}$ by the lemma, and $t^{-1}(o) = t^{-1}(N'^0) = (t'(N'))^0$, which is $\sigma(E, E')$-closed, being the polar of a subset of E'. Thus $t^{-1}(o)$ is closed.

The definition of a nearly open mapping suggests a parallel definition of a nearly continuous mapping. If t is a linear mapping of the convex space E into the convex space F, we shall call t *nearly continuous* if $\overline{t^{-1}(V)}$ is a neighbourhood in E for every neighbourhood V in F. When t is (1, 1) and onto, clearly t is nearly continuous if and only if its inverse is nearly open.

LEMMA 5. *Let E and F be separated convex spaces with duals E′ and F′. If t is a nearly continuous linear mapping of E into F, with transpose t′ (mapping F′ into E*), and if M′ is a nearly closed vector subspace of E′, then $t'^{-1}(M')$ is nearly closed.*

Proof. Let $N' = t'^{-1}(M')$; we have to show that $N' \cap V^0$ is $\sigma(F', F)$-closed for each neighbourhood V in F. Let $W = t^{-1}(V)$. Then $\overline{W}$ is a neighbourhood in E since t is nearly continuous; therefore, since M' is nearly closed, $M' \cap \overline{W}^0$ is $\sigma(E^*, E)$-compact and so $\sigma(E^*, E)$-closed. Hence $t'^{-1}(M' \cap \overline{W}^0)$ is $\sigma(F', F)$-closed. Now

$$t'^{-1}(M' \cap \overline{W}^0) = t'^{-1}(M' \cap W^0) = N' \cap t'^{-1}(W^0)$$
$$= N' \cap (t(W))^0 = N' \cap (V \cap t(E))^0.$$

Hence

$$N' \cap V^0 = N' \cap V^0 \cap (V \cap t(E))^0 = t'^{-1}(M' \cap \overline{W}^0) \cap V^0,$$

an intersection of $\sigma(F', F)$-closed sets.

PROPOSITION 10. *Let E and F be separated convex spaces and let t be a linear mapping of E into F with a closed graph.*

(i) *If t is nearly continuous and F is fully complete, then t is continuous.*

(ii) *If t is nearly open and onto and E is fully complete, then t is open.*

Proof. (i) We prove first that t is weakly continuous, by showing that $t'^{-1}(E') = F'$, so that $t'(F') \subseteq E'$. By Lemma 5, with $M' = E'$, $t'^{-1}(E')$ is nearly closed; since F is fully complete, it is weakly closed. But by Lemma 4, $t'^{-1}(E')$ is dense in F' since the graph of t is closed. Hence $t'^{-1}(E') = F'$, and so t is weakly continuous. Now if V is a closed absolutely convex neighbourhood in F, V is also weakly closed; therefore $t^{-1}(V)$ is weakly closed and so closed. But $\overline{t^{-1}(V)}$ is a neighbourhood since t is nearly continuous; thus $t^{-1}(V)$ is a neighbourhood, and t is proved continuous.

(ii) By the Corollary to Lemma 4, $t^{-1}(o)$ is closed and so the space $E/t^{-1}(o)$, under the quotient topology, is separated. Then we can write $t = s \circ k$, where k is the canonical mapping of E onto $E/t^{-1}(o)$ and s is a $(1,1)$ linear mapping of $E/t^{-1}(o)$ onto F. If U is a neighbourhood in E, then $\overline{s(k(U))} = \overline{t(U)}$ and so s is nearly open since t is. If M' is the dual of $E/t^{-1}(o)$, then $s'^{-1}(M') = t'^{-1}(E')$, since k' is the injection mapping of M' into E', and so by Lemma 4 applied twice, the graph of s is closed. Hence s^{-1} is nearly continuous, has a closed graph, and maps F into $E/t^{-1}(o)$, which is fully complete by the Corollary of Proposition 9. Therefore, by (i), s^{-1} is continuous. Thus s, and so also t, is open.

THEOREM 6. (Closed graph theorem.) *A linear mapping, with a closed graph, of a separated barrelled space into a fully complete space is continuous.*

Proof. If t is such a mapping and V is an absolutely convex neighbourhood in the fully complete space, then $\overline{t^{-1}(V)}$ is a barrel and so a neighbourhood. Thus t is nearly continuous and the theorem follows from Proposition 10 (i).

THEOREM 7. (Open mapping theorem.) *A continuous linear mapping of a fully complete space onto a separated barrelled space is open; more generally, a linear mapping, with a closed graph, of a fully complete space onto a separated barrelled space is open.*

Proof. This follows from the second part of Proposition 10 in the same way that Theorem 6 follows from the first part.

COROLLARY 1. *If a separated barrelled space is the continuous image of a fully complete space, it is fully complete.*

By the Theorem and Proposition 9.

COROLLARY 2. *A continuous* $(1,1)$ *linear mapping of a fully complete space onto a separated barrelled space is an isomorphism.*

Corollary 3. *A fully complete space cannot have a strictly coarser separated barrelled topology.*

By Corollary 2, applied to the identity mapping.

We conclude this section by demonstrating senses in which Theorems 6 and 7 contain the 'best possible' conditions on the spaces, that one be barrelled and the other fully complete, if reasonable closed graph and open mapping theorems are required to hold. First we have:

Proposition 11. *Let E be a separated convex space and suppose that every linear mapping, with a closed graph, of E into any Banach space is continuous. Then E is barrelled.*

Proof. Let B be a barrel in E and let

$$N = \bigcap_{\epsilon > 0} \epsilon B.$$

Let k be the canonical mapping of E onto E/N with the norm topology for which $k(B)$ is the unit ball; let F be the completion of E/N with this topology, and denote by V the unit ball in F, so that $V = \overline{k(B)}$. We show that the graph G of k in $E \times F$ is closed. Suppose that $(x, y) \in \overline{G}$. We have to prove that

$$y - k(x) = o.$$

Since E/N is dense in F, for each $\epsilon > 0$ there is a point $z \in E$ with $k(z) \in y - k(x) + \epsilon V$. Also for each absolutely convex neighbourhood U in E, $(x + U, y + \epsilon V) \cap G \neq \emptyset$ and so $y - k(x) \in k(U) + \epsilon V$. Hence $k(z) \in k(U) + 2\epsilon V$, and, since $(E/N) \cap V = k(B)$,

$$z \in U + 2\epsilon B + N = U + 2\epsilon B,$$

for all absolutely convex neighbourhoods U. Thus

$$z \in 2\epsilon \overline{B} = 2\epsilon B,$$

so that $y - k(x) \in 2\epsilon k(B) + \epsilon V \subseteq 3\epsilon V$. Hence $y = k(x)$ and the graph of k is closed. By hypothesis, k is therefore continuous and so $k^{-1}(V) = k^{-1}(k(B)) = B$ is a neighbourhood in E.

This result makes it clear why we have to make one space barrelled in Theorems 6 and 7. An exactly parallel result justifying the hypothesis of full completeness might be expected to assert, roughly, that if we have a closed graph theorem for mappings from Banach spaces into F then F is fully complete, but this is unfortunately false (see Suppl. 2). However, we can show that in Proposition 10 the condition of full completeness cannot be relaxed; if E is a separated convex space with the property that every nearly open mapping, with a closed graph, of E onto any separated convex space is open, then E must be fully complete. (Alternatively, it is also sufficient that any nearly continuous mapping, with a closed graph, of any separated convex space into any separated quotient of E should be continuous.) We prove a simpler, and slightly stronger, form of converse:

PROPOSITION 12. *Let E be a separated convex space. If every continuous nearly open linear mapping of E onto any separated convex space is open, then E is fully complete.*

Proof. Let M' be a nearly closed vector subspace of the dual E' of E, and consider the canonical mapping k of E onto $F = E/M'^0$, under the topology of uniform convergence on the sets $M' \cap U^0$, as U runs through a base $\mathscr{U}$ of absolutely convex neighbourhoods in E. First, the dual of F under this topology is M'. For, if $U \in \mathscr{U}$, then $M' \cap U^0$ is $\sigma(E', E)$-compact and so certainly $\sigma(M', F)$-compact; thus the topology is one of uniform convergence on $\sigma(M', F)$-compact sets and so (Ch. III, Prop. 8) the dual is M'. Next, k is nearly open. For

$$\overline{k(U)} = (k(U))^{00} = (k'^{-1}(U^0))^0 = (M' \cap U^0)^0,$$

k' being the canonical mapping of M' into E', and so each $\overline{k(U)}$ is a neighbourhood in F. Also k is continuous. For the sets $(M' \cap U^0)^0$ $(U \in \mathscr{U})$ form a base of neighbourhoods in F, and, since $k(U) \subseteq \overline{k(U)} = (M' \cap U^0)^0$, the quotient topology is finer than the topology on F. By hypothesis, k is therefore open. Hence the topology coincides with the quotient topology, under which F has dual M'^{00}. Therefore $M' = M'^{00}$, that is, M' is $\sigma(E', E)$-closed; thus E is fully complete.

3. Fréchet spaces. The theory of Fréchet spaces, including as it does as a special case the theory of Banach spaces, is too vast a subject to be dealt with at all adequately in a volume of this size and scope. All we shall attempt to do in this section is to collect together some of the main properties of Fréchet spaces, especially those depending upon the ideas that we have developed in earlier sections.

In the first place, a Fréchet space (a complete metrisable convex space) is barrelled (Ch. IV, Th. 2). Hence if E is a Fréchet space and F is any convex space, every pointwise bounded set of continuous linear mappings of E into F is equicontinuous, and in particular every pointwise convergent sequence of continuous linear mappings of E into F converges to a continuous linear mapping, uniformly on each compact subset of E (Banach–Steinhaus theorem, Ch. IV, Th. 3). A Fréchet space E with dual E' has the topology $\tau(E, E')$ (Ch. IV, Th. 1, Corollary) and therefore every weakly continuous linear mapping of E into another separated convex space is continuous under the initial topologies (Ch. III, Prop. 14). More generally, since a Fréchet space E is a Mackey space (Ch. V, Prop. 8) every bounded linear mapping defined on E is continuous.

A closed vector subspace of a Fréchet space is a Fréchet space; the next proposition shows that the same is true for a continuous open image in a separated space and in particular for a separated quotient. (We could deduce this from Theorem 5 and Proposition 9 but the direct proof is much simpler.)

Proposition 13. *If there is a continuous open linear mapping of a Fréchet space onto a separated convex space F, then F is a Fréchet space.*

Proof. Let (U_n) be a base of absolutely convex neighbourhoods in the Fréchet space E, with $U_{n+1} \subseteq \frac{1}{2}U_n$ for each n. If t is a continuous open linear mapping of E onto F, the sets $t(U_n)$ form a base of neighbourhoods in F and so F is metrisable. We have to prove F complete. Suppose that (y_n) is a Cauchy sequence in F. There is a subsequence $(y_{n(r)})$ with $y_{n(r+1)} - y_{n(r)} \in t(U_r)$ for each r. We can now choose $x_{n(1)}, x_{n(2)}, \ldots$, successively so that

$$t(x_{n(r)}) = y_{n(r)} \quad \text{and} \quad x_{n(r+1)} - x_{n(r)} \in U_r.$$

Then, if $r < s$,

$$x_{n(s)} - x_{n(r)} = \sum_{r \leqslant i \leqslant s-1} (x_{n(i+1)} - x_{n(i)}) \in \sum_{r \leqslant i \leqslant s-1} U_i \subseteq U_{r-1},$$

and so $(x_{n(r)})$ is a Cauchy sequence in E. Since E is complete, $x_{n(r)} \to a$, say, and then $y_{n(r)} = t(x_{n(r)}) \to t(a)$. Hence $y_n \to t(a)$ and thus F is complete.

COROLLARY. *A quotient of a Fréchet space by a closed vector subspace is a Fréchet space.*

For if M is a closed vector subspace of the Fréchet space E, then E/M is separated and the canonical mapping of E onto E/M is continuous and open.

In § 2 we proved (Th. 5) that every Fréchet space is fully complete. Since every Fréchet space is also barrelled, Theorems 6 and 7 give immediately:

THEOREM 8. *A linear mapping, with a closed graph, of one Fréchet space into another is continuous. A continuous linear mapping of one Fréchet space onto another is open.*

COROLLARY 1. *A* (1, 1) *continuous linear mapping of one Fréchet space onto another is an isomorphism. In particular, a vector space cannot have two distinct comparable topologies under each of which it is a Fréchet space.*

COROLLARY 2. *If a Fréchet space is the algebraic direct sum of two closed vector subspaces, it is their topological direct sum.*

Proof. Each of the subspaces is a Fréchet space and therefore their sum, under the direct sum topology, is complete (Ch. V, Prop. 23; or, indirectly but more simply, by Ch. V, Prop. 21 and Prop. 18, Cor. 1) and is clearly metrisable. Hence, by Corollary 1, the direct sum topology and the (necessarily coarser) original topology coincide.

PROPOSITION 14. *If E and F are Fréchet spaces and t is a continuous linear mapping of E into F, then t is an open mapping of E onto $t(E)$ if and only if $t(E)$ is closed in F.*

Proof. If $t(E)$ is closed in F, $t(E)$ is a Fréchet space and so t is open (Th. 8). Conversely, if t is an open mapping of E onto $t(E)$, then $t(E)$ is a Fréchet space (Prop. 13) and so $t(E)$ is closed in F.

Since every Fréchet space E is a Mackey space, its dual E' is complete under any topology between the topology κ of uniform convergence on compact sets and the strong topology $\beta(E', E)$

(Prop. 1). In this case, it follows from the Mackey–Arens theorem (Ch. III, Th. 7) that κ is a topology of the dual pair (E', E) because E is complete (Ch. III, Th. 5, Corollary), and so this range of topologies includes $\tau(E', E)$. (In fact, E' is fully complete under any topology between κ and $\tau(E', E)$; see Suppl. 1.)

In the special case when E is a Banach space, its strong dual is also a Banach space; the next proposition shows that this is the only case in which the strong dual of a Fréchet space can be a Fréchet space. In fact we prove, more generally:

PROPOSITION 15. *If the strong dual of a metrisable convex space E is also metrisable, then E is normable.*

Proof. Let (U_n) be a base of neighbourhoods in E with $U_{n+1} \subseteq U_n$ for each n, and let (A_n) be a sequence of closed absolutely convex bounded sets whose polars form a base of neighbourhoods in E' under $\beta(E', E)$. We prove that, for some n, $U_n \subseteq A_n$; by Chapter III, Theorem 1, it will follow that E is normable. Suppose that, on the contrary, there are points $x_n \in U_n$ with $x_n \notin A_n$. Then the set B of all the points x_n is bounded and therefore B^0 is a $\beta(E', E)$-neighbourhood. Hence there is an n with $A_n^0 \subseteq B^0$ and so $B \subseteq A_n^{00} = A_n$, which contradicts the definition of B.

Although the strong dual of a Fréchet space E is, in general, not metrisable, the bidual E'' under its strong topology is. For this topology coincides with the topology of uniform convergence on equicontinuous sets, E being a Mackey space (Ch. V, Prop. 9) and so the bipolars in E'' of a countable neighbourhood base in E form a countable neighbourhood base in E''. We can go further:

PROPOSITION 16. *The bidual of a metrisable space is a Fréchet space under its strong topology.*

Proof. Let (U_n) be a base of closed absolutely convex neighbourhoods in the metrisable space E, with $U_{n+1} \subseteq \frac{1}{2}U_n$ for each n. If (x''_n) is any Cauchy sequence in the bidual E'', there is a subsequence (which we may suppose for convenience to be the sequence (x''_n) itself) for which $x''_{n+1} - x''_n \in \frac{1}{2}U_n^{00}$ for each n. Also there are closed absolutely convex bounded subsets A_n of E with

$$x''_{n+1} - x''_n \in \tfrac{1}{2}A_n^{00}.$$

Now it follows from the definition of a polar that

$$\tfrac{1}{2}(A_n^{00} \cap U_n^{00}) \subseteq (A_n^0 + U_n^0)^0.$$

Further, $A_n^0 + U_n^0$ is $\sigma(E', E)$-closed, being the sum of a weakly closed and a weakly compact set (Ch. III, Lemma 7), and so it contains the polar of $A_n \cap U_n$. Thus

$$x''_{n+1} - x''_n \in \tfrac{1}{2}(A_n^{00} \cap U_n^{00}) \subseteq (A_n^0 + U_n^0)^0 \subseteq (A_n \cap U_n)^{00}$$

and so

$$x''_{n+1} \in x''_1 + \sum_{1 \leqslant r \leqslant n} (A_r \cap U_r)^{00} \subseteq x''_1 + (\sum_{1 \leqslant r \leqslant n} A_r \cap U_r)^{00},$$

since any point of $\sum_{1 \leqslant r \leqslant n} (A_r \cap U_r)^{00}$ belongs to the $\sigma(E'', E')$-closure of $\sum_{1 \leqslant r \leqslant n} (A_r \cap U_r)$. Now let

$$A = \bigcup_{n=1}^{\infty} \sum_{1 \leqslant r \leqslant n} (A_r \cap U_r);$$

then $x''_{n+1} \in x''_1 + A^{00}$. We prove that A is bounded. If U is any neighbourhood in E, there is an m with $U_m \subseteq U$ and then for $n > m$

$$\sum_{1 \leqslant r \leqslant n} (A_r \cap U_r) \subseteq \sum_{1 \leqslant r \leqslant m} A_r + \sum_{m < r \leqslant n} U_r \subseteq \sum_{1 \leqslant r \leqslant m} A_r + U_m \subseteq \lambda U$$

for some $\lambda > 0$. Also

$$\bigcup_{1 \leqslant n \leqslant m} \sum_{1 \leqslant r \leqslant n} (A_r \cap U_r)$$

is bounded and therefore so is A. It follows that A^{00} is $\sigma(E'', E')$-compact, and so $\beta(E'', E')$-complete (Prop. 3, Corollary, applied to the dual pair (E', E'')). Hence (x''_n) converges in A^{00}, and thus E'' is complete.

SUPPLEMENT

(1) *Fully complete spaces.* We give first the example promised of a space that is complete but not fully complete. Let E be an infinite dimensional Banach space, and consider E under the finest convex topology $\tau(E, E^*)$. Then E is complete (Ch. III, Suppl. 2) but it follows from Theorem 7 (the open mapping

theorem) that E cannot be fully complete. For the identity mapping of E under $\tau(E, E^*)$ onto E under its Banach space topology is continuous but cannot be open, since an infinite dimensional normed space cannot have the topology $\tau(E, E^*)$ (Ch. III, Suppl. 2) and a Banach space is barrelled.

The quotient of a fully complete space by a closed vector subspace is also fully complete (Prop. 9, Corollary); however, the quotient of a complete space by a closed vector subspace need not be complete. Grothendieck gives an example ('Sur les espaces (F) et (DF)', *Summa Brasil. Math.* 3 (1954), 57–121, II, 2, 1°) of a complete space, which is moreover a strict inductive limit of a sequence of Fréchet spaces (see Ch. VII, § 1), with an incomplete quotient. Thus a strict inductive limit of a sequence of fully complete spaces may fail to be fully complete (even when they are Fréchet spaces), another point of difference from the corresponding completeness property (see Ch. VII, Prop. 3).

The space E^* under the topology $\sigma(E^*, E)$ is fully complete, since every vector subspace of E is $\sigma(E, E^*)$-closed.

The dual of a Fréchet space is fully complete under any topology between the topology of uniform convergence on compact sets and $\tau(E', E)$. (For under such a topology E' has dual E. If M is a nearly closed vector subspace of E, every point $a \in \overline{M}$ is the limit of a convergent sequence (x_n) of points of M. Now the set A consisting of the points x_n and the point a is compact, and so A^{00} is equicontinuous. Hence $M \cap A^{00}$ is closed and therefore $a \in M$. Thus M is closed.) In particular, the strong dual of a reflexive Fréchet space is fully complete.

(2) *Closed graph and open mapping theorems.* For Fréchet spaces, the open mapping theorem can be proved more directly by the use of category arguments, and the closed graph theorem deduced immediately from it. (See, for example, Banach, *Théorie des opérations linéaires* (1932), Ch. III, § 3, Ths. 3–7: in fact the spaces need only be complete metric vector spaces and not necessarily locally convex.) The proof of Theorem 5 above, that a Fréchet space is fully complete, is similar to part of Banach's open mapping theorem. The third theorem of Banach mentioned at the beginning of § 2, proving (somewhat in disguise) that a Banach space is fully complete, is Lemma 3 of Chapter VIII

of his book. The connection between the open mapping theorem and full completeness for convex spaces was noticed by Pták ('On complete topological linear spaces', *Czechoslovak Math. J.* 3 (78) (1953), 301–64). There have been further generalisations; in particular, Dieudonné and Schwartz use a category method and Banach's theorems to prove that a continuous linear mapping of one strict inductive limit of a sequence of Fréchet spaces onto another is open ('La dualité dans les espaces $(\mathscr{F})$ et $(\mathscr{L}\mathscr{F})$', *Ann. Inst. Fourier Grenoble*, 1 (1950), 61–101, Théorème 1). More generally, we have shown (*Proc. Glasgow Math. Assoc.* 3 (1956), 9–12) that if E is a separated inductive limit of convex Baire spaces and F a separated inductive limit of a sequence of fully complete spaces, the union of whose images is F, then any linear mapping, with a closed graph, of E into F is continuous, and any continuous linear mapping of F onto E is open.

This last theorem can be used to demonstrate the failure of the exact parallel to Proposition 11 mentioned in the text. We have to show that a separated convex space F need not be fully complete even if every linear mapping, with a closed graph, from any Banach space into F is continuous. Now if F is a strict inductive limit of a sequence of Fréchet spaces then certainly, by this theorem, the hypothesis is satisfied, but Grothendieck's example (see Suppl. 1) shows that there is such a space F that is not fully complete. The true situation is somewhat untidy. If we suppose that a closed graph theorem holds from any separated barrelled space not only into F but into any separated quotient of F, then F must be fully complete, provided that F itself is also supposed to be barrelled. This is simpler in the form of an open mapping theorem: a separated *barrelled* space must be fully complete if every continuous linear mapping of it onto any separated barrelled space is open. (The proof is similar to that of Proposition 12; from the fact that the space is barrelled, it follows that its quotient, under the topology used in that proof, is also barrelled.) In conjunction with Grothendieck's example again, this last result shows that, even if F is the inductive limit of a sequence of Fréchet spaces, it is not possible to improve the closed graph and open mapping theorems quoted at the end of the previous paragraph by supposing merely that E be barrelled.

Another category result proved by Banach (*Théorie des opérations linéaires* (1932), Ch. III, § 3, Th. 3) is that if E and F are complete metric spaces and t is a continuous linear mapping of E into F, then $t(E)$ is either of the first category in F or is the whole of F. This is also true when E is fully complete or the inductive limit of a sequence of fully complete spaces, the union of whose images is E, and F is any separated convex space. (For in the first case, if $t(E)$ is of the second category in F then it is barrelled (Ch. IV, Suppl. 1) and so fully complete (Th. 7, Cor. 1); thus closed in F. But $t(E)$ is dense in F (Ch. IV, Suppl. 1) and so $t(E) = F$. In the second case, if E is the inductive limit of the sequence (E_n) by the mappings (u_n), then

$$t(E) = \bigcup_{n=1}^{\infty} t(u_n(E_n)).$$

Hence if $t(E)$ is of the second category in F, there is some n with $t(u_n(E_n))$ of the second category in F. By the same argument as before, applied to the mapping $t \circ u_n$, we have $t(u_n(E_n)) = F$ and so $t(E) = F$.)

(3) *Hypercomplete spaces.* The proof that a Fréchet space is fully complete (Th. 5) may be adapted to show that a nearly closed absolutely convex subset of the dual is weakly closed. (For suppose that M' is absolutely convex and nearly closed, and that $x' \in M'^{00}$. Then $x' \in U^0$ for some neighbourhood U in E. If $\epsilon > 0$ then $(M' \cap (\frac{1}{4}\epsilon U)^0)^0 \subseteq M'^0 + \epsilon U$ [for the part of the proof that shows the existence of a suitable $V = \frac{1}{4}U$ is not dependent on the fact that, in Theorem 5, M' is a vector subspace]. Hence if $x \in (M' \cap (\frac{1}{4}\epsilon U)^0)^0$, then $|\langle x, x'\rangle| \leqslant 1+\epsilon$ and so

$$x' \in (1+\epsilon)(M' \cap (\tfrac{1}{4}\epsilon U)^0)^{00} = (1+\epsilon)(M' \cap (\tfrac{1}{4}\epsilon U)^0) \subseteq (1+\epsilon) M'.$$

Also $x' \in U^0 \subseteq (1+\epsilon) U^0$ and so $x' \in (1+\epsilon)(M' \cap U^0)$ for every $\epsilon > 0$. Since $M' \cap U^0$ is weakly compact, $x' \in M' \cap U^0 \subseteq M'$.) Kelley has called a separated convex space *hypercomplete* if it has the property that nearly closed absolutely convex subsets of the dual are weakly closed; a space is hypercomplete if and only if the space of its absolutely convex subsets, under the Hausdorff uniform structure, is complete. (See J. L. Kelley, 'Hypercomplete linear topological spaces', *Michigan Math. J.* 5 (1958), 235–46.)

(4) *Krein's theorem.* There are several theorems relating to weakly compact sets in a convex space, of which perhaps the most striking is one due to M. Krein: in a complete separated convex space, the closed absolutely convex envelope of every weakly compact set is weakly compact. One method of proving this uses Eberlein's theorem (Th. 4) and also Lebesgue's theorem of dominated convergence (see e.g. Grothendieck, *Espaces vectoriels topologiques* [São Paulo, 1954], Ch. v, § 3, Th. 4, or Bourbaki, *Éléments de mathématique.* Livre VI. *Intégration,* Ch. IV, § 4, ex. 19).

CHAPTER VII

SOME FURTHER TOPICS

1. Strict inductive limits. As well as Fréchet spaces themselves, inductive limits of Fréchet spaces arise frequently in applications. In practice these inductive limits are of a rather special type, and consequently they have certain additional properties not possessed by general inductive limits of Fréchet spaces. Some of these properties do not depend on the fact that the defining spaces are Fréchet spaces and so we study these special inductive limits without any initial restriction on the defining spaces.

Suppose that E is a vector space and that (E_n) is a strictly increasing sequence of vector subspaces whose union is E. Suppose also that each E_n has a topology ξ_n under which it is a convex space, and that, for each n, the topology induced on E_n by the topology ξ_{n+1} on E_{n+1} is ξ_n. Then each E_n is embedded, algebraically and topologically, in E_{n+1} (and therefore in E_{n+r} for each r). Let ξ be the inductive limit topology on E, so that ξ is the finest convex topology on E inducing on each E_n a topology coarser than ξ_n. Then E with the convex topology ξ is called the *strict inductive limit* of the vector subspaces E_n. In this case we can show that the topology on E induces exactly ξ_n on each E_n.

PROPOSITION 1. *Let the convex space E with topology ξ be the strict inductive limit of the convex spaces E_n, with topologies ξ_n. Then ξ induces ξ_n on each E_n.*

Proof. Let U_n be any absolutely convex neighbourhood in E_n under ξ_n. We prove that there is a ξ-neighbourhood U in E with $U \cap E_n = U_n$, thus showing that ξ induces on E_n a topology finer than, and therefore identical with, ξ_n. Since ξ_{n+1} induces ξ_n on E_n, there is an absolutely convex ξ_{n+1}-neighbourhood U_{n+1} with $U_{n+1} \cap E_n \subseteq U_n$. Continuing in this way, we can define for each r an absolutely convex ξ_{n+r}-neighbourhood U_{n+r} such that $U_{n+r} \cap E_{n+s} \subseteq U_{n+s}$ for $0 \leqslant s \leqslant r$. Let U be the absolutely convex envelope of $\bigcup_{r=0}^{\infty} U_{n+r}$; then it is easy to verify that $U_{n+r} \subseteq U \cap E_{n+r}$ for $r \geqslant 0$. Also for $m \leqslant n$, $U \cap E_m = U_n \cap E_m$ and so U is a ξ-neighbourhood with the required property.

Proposition 2. *The strict inductive limit of a sequence of separated convex spaces is separated.*

Proof. If E is the strict inductive limit of the separated convex spaces E_n and if $x \neq o$, then there is some n with $x \in E_n$. There is therefore a neighbourhood U_n in E_n with $x \notin U_n$. By Proposition 1 there is a neighbourhood U in E with $U \cap E_n \subseteq U_n$ and so $x \notin U$. Thus E is separated.

Proposition 3. *The strict inductive limit of a sequence of complete separated convex spaces is complete.*

Proof. If the strict inductive limit E of the sequence of complete separated convex spaces E_n is not complete, there is a point z in its completion $\hat{E}$ but not in any E_n. Since each E_n is closed in $\hat{E}$, there are absolutely convex neighbourhoods W_n in $\hat{E}$ with $(z + W_n) \cap E_n = \emptyset$, and these can be chosen so that $W_{n+1} \subseteq W_n$ for each n. Let U be the absolutely convex envelope of

$$\bigcup_{n=1}^{\infty} (\tfrac{1}{2}W_n \cap E_n).$$

Then U is a neighbourhood in E and so its closure $\overline{U}$ is a neighbourhood in $\hat{E}$. Since E is dense in $\hat{E}$, $z + \overline{U}$ meets E and therefore meets some E_n. We show below that $\overline{U} \subseteq W_n + E_n$; this will imply that $z + W_n$ meets E_n, which is a contradiction and thus establishes the proposition.

It remains to prove that $\overline{U} \subseteq W_n + E_n$. Now $\overline{U} \subseteq U + \frac{1}{2}W_n$, and any element of U is of the form $\sum_{1 \leqslant r \leqslant s} \lambda_r x_r$ with $x_r \in \frac{1}{2}W_r \cap E_r$ and $\sum_{1 \leqslant r \leqslant s} |\lambda_r| = 1$. Then

$$\sum_{r \leqslant n} \lambda_r x_r \in \sum_{r \leqslant n} \lambda_r E_r \subseteq E_n \quad \text{and} \quad \sum_{r > n} \lambda_r x_r \in \sum_{r > n} \lambda_r . \tfrac{1}{2}W_r \subseteq \tfrac{1}{2}W_n,$$

and so $U \subseteq \frac{1}{2}W_n + E_n$. Hence $\overline{U} \subseteq \frac{1}{2}W_n + E_n + \frac{1}{2}W_n = W_n + E_n$.

When each defining space E_n is complete, the bounded sets of the strict inductive limit have a simple characterisation.

PROPOSITION 4. *Suppose that E is the strict inductive limit of the convex spaces E_n, and that for each n, E_n is a closed vector subspace of E_{n+1}. Then a subset of E is bounded if and only if it is contained in and bounded in one of the E_n.*

Proof. Let A be a subset of E not contained in any E_n. Then there is a sequence $(n(r))$ of positive integers and a sequence (x_r) of points with $x_r \notin E_{n(r)}$, $x_r \in A$ and $x_r \in E_{n(r+1)}$. Since each $E_{n(r)}$ is closed in $E_{n(r+1)}$, we can define successively absolutely convex $\xi_{n(r)}$-neighbourhoods $U_{n(r)}$ such that $(1/r)\,x_r + U_{n(r+1)}$ does not meet $E_{n(r)}$ and $U_{n(r+1)} \cap E_{n(r)} \subseteq U_{n(r)}$. Then if U is the absolutely convex envelope in E of $\bigcup_{r=1}^{\infty} U_{n(r)}$, U is a ξ-neighbourhood and $U \cap E_{n(r+1)} \subseteq E_{n(r)} + U_{n(r+1)}$, so that U does not contain $(1/r)\,x_r$ for any r. Thus U does not absorb A and therefore A is unbounded in E. Hence a bounded subset of E must be contained in one of the E_n (and clearly it must be bounded in this E_n). Conversely, such a set is bounded in E.

The condition that each E_n is closed in E_{n+1} is satisfied when the E_n are complete. This condition also ensures that E is not metrisable even if each E_n is a Fréchet space:

PROPOSITION 5. *Suppose that E is the strict inductive limit of the convex spaces E_n and that, for each n, E_n is a closed vector subspace of E_{n+1}. Then E is not metrisable.*

Proof. If (U_n) is any decreasing sequence of neighbourhoods in E, there are points $x_n \in U_n$ with $x_n \notin E_n$. Then by Proposition 4, the set of points x_n cannot be bounded. But it is absorbed by each U_n and so (U_n) cannot be a neighbourhood base. Hence E is not metrisable.

We conclude this section by collecting together some of the properties of strict inductive limits of Fréchet spaces. Such a space is separated (Prop. 2), complete (Prop. 3) but not metrisable (Prop. 5), barrelled (Ch. v, Prop. 6) and a Mackey space (Ch. v, Prop. 8, Corollary). If E is such a space and (E_n) the defining sequence of Fréchet spaces, then the topology of each E_n coincides with that induced on it by that of E (Prop. 1) and a subset A of E is bounded if and only if it is contained in and bounded in some E_n (Prop. 4).

2. Bilinear mappings and tensor products. The main purpose of this section is to describe a valuable technical device by which the study of bilinear mappings and forms can be reduced to that of linear mappings and forms. Suppose that E, F and G are three vector spaces over the same field and that h is a mapping of $E \times F$ into G. Then, for each fixed $y \in F$, h defines a partial mapping h_y of E into G, obtained by putting

$$h_y(x) = h(x, y),$$

and, similarly, h defines a partial mapping h_x of F into G for each fixed $x \in E$. If all the partial mappings h_x and h_y are linear, h is called a *bilinear mapping* of $E \times F$ into G. When G is the scalar field, h is called a *bilinear form* on $E \times F$. The device mentioned above is to construct a vector space, called the tensor product of E and F, with the property that the bilinear mappings of $E \times F$ into G correspond exactly to the linear mappings of this tensor product into G. There are various methods of introducing this tensor product; the one that we adopt so constructs the tensor product that the bilinear forms have the required correspondence. We then prove that it extends to bilinear mappings.

Denote by T^* the set of all bilinear forms on $E \times F$; then, with the obvious definitions of addition and multiplication by scalars, T^* is a vector space. Let T^{**} be its algebraic dual. There is a natural mapping ϕ of $E \times F$ into T^{**}, obtained by putting

$$(\phi(x, y))(h) = h(x, y)$$

for $(x, y) \in E \times F$ and $h \in T^*$. This mapping ϕ is clearly bilinear. The image $\phi(E, F)$ of $E \times F$ by ϕ is not in general a vector subspace of T^{**}, though it is easy to see that if $w \in \phi(E, F)$, then $\lambda w \in \phi(E, F)$ for every scalar λ. The vector subspace of T^{**} spanned by $\phi(E, F)$ is called the *tensor product* of E and F and denoted by $E \otimes F$. We denote $\phi(x, y)$ by $x \otimes y$; if $a \neq o$ and $b \neq o$, then $a \otimes b \neq o$. (Take $a^* \in E^*$ with $\langle a, a^* \rangle \neq 0$ and $b^* \in F^*$ with $\langle b, b^* \rangle \neq 0$, and then put $h(x, y) = \langle x, a^* \rangle \langle y, b^* \rangle$; then $h \in T^*$ and $h(a, b) \neq 0$.) Every element of $E \otimes F$ can be written as a finite sum of the form $\sum_i x_i \otimes y_i$ with $x_i \in E$ and $y_i \in F$, but, in general, this representation is by no means unique.

Now $(T^*, E\otimes F)$ is a dual pair, since, for $h \in T^*$, $h \neq o$ means that there are elements $x \in E$, $y \in F$ with $h(x,y) \neq 0$, and this is equivalent to $\langle h, x\otimes y\rangle \neq 0$. Moreover, T^* is the algebraic dual of $E\otimes F$, because to any linear form f on $E\otimes F$ corresponds the element $f\circ\phi$ of T^*. Thus the tensor product $T = E\otimes F$ has the property that its algebraic dual T^* is the set of all bilinear forms on $E\times F$. This result extends to bilinear mappings:

Proposition 6. *Let E, F and G be three vector spaces over the same scalar field and let ϕ be the canonical mapping of $E\times F$ into $E\otimes F$. Then to every linear mapping f of $E\otimes F$ into G corresponds a bilinear mapping $f\circ\phi$ of $E\times F$ into G, and this correspondence is an isomorphism of the vector space of linear mappings of $E\otimes F$ into G onto the vector space of bilinear mappings of $E\times F$ into G.*

Proof. Denoting this correspondence by t, so that $t(f) = f\circ\phi$, it is clear that t is linear and $(1,1)$; we must prove that every bilinear mapping h is an image by t. Now for each $z^* \in G^*$, we can define a bilinear form $f^*(z^*)$ on $E\times F$, by putting

$$(f^*(z^*))(x,y) = \langle h(x,y), z^*\rangle.$$

Then f^* is a linear mapping of G^* into T^*; let f be its transpose, restricted to $E\otimes F$. We have

$$\langle f(\phi(x,y)), z^*\rangle = \langle \phi(x,y), f^*(z^*)\rangle = \langle h(x,y), z^*\rangle$$

for all $(x,y)\in E\times F$ and all $z^*\in G^*$. Hence f maps $E\otimes F$ into G and $h = f\circ\phi = t(f)$.

When E and F are convex spaces, $E\otimes F$ can be given a topology in such a way that the correspondence between linear mappings f and bilinear mappings $f\circ\phi$ preserves continuity.

Proposition 7. *Let E and F be two convex spaces over the same field and let ϕ be the canonical mapping of $E\times F$ into $E\otimes F$. There is a finest convex topology on $E\otimes F$ under which ϕ is continuous. If $\mathscr{U}$, $\mathscr{V}$ are neighbourhood bases in E, F, the absolutely convex envelopes of the sets $\phi(U,V)$ $(U\in\mathscr{U}, V\in\mathscr{V})$ form a neighbourhood base for this topology. A linear mapping f of $E\otimes F$ under this topology into a third convex space G is continuous if and only if the bilinear mapping $f\circ\phi$ of $E\times F$ into G is continuous.*

Proof. Let W be an absolutely convex neighbourhood in any topology on $E \otimes F$ under which ϕ is continuous. Then $\phi^{-1}(W)$ contains a set $U \times V$ with $U \in \mathscr{U}$ and $V \in \mathscr{V}$ and so W contains the absolutely convex envelope of $\phi(U, V)$. As U and V run through $\mathscr{U}$ and $\mathscr{V}$, these sets form a base of neighbourhoods for a convex topology on $E \otimes F$; it is therefore the finest convex topology making ϕ continuous. It is now easily verified that f is continuous if and only if $f \circ \phi$ is continuous at the origin. But continuity at the origin implies continuity everywhere for a bilinear mapping (the proof of this is similar to that for a linear mapping; cf. Ch. II, Prop. 1).

We shall call the space $E \otimes F$ with this topology the *topological tensor product* of the convex spaces E and F. (There are other useful topologies on $E \otimes F$, of which we shall give one example later; the topology of Proposition 7 is sometimes called the *projective* tensor product topology.)

In order to deal with the separatedness of this topology, we need an algebraic result.

LEMMA 1. *If (E, E') and (F, F') are dual pairs, so is $(E \otimes F, E' \otimes F')$, with the bilinear form*

$$\langle w, w' \rangle = \sum_{i,j} \langle x_i, x'_j \rangle \langle y_i, y'_j \rangle,$$

where

$$w = \sum_i x_i \otimes y_i \in E \otimes F \quad \textit{and} \quad w' = \sum_j x'_j \otimes y'_j \in E' \otimes F'.$$

Proof. Let $w = \sum_i x_i \otimes y_i$ be any element of $E \otimes F$ other than the origin. Take a base $e_1, e_2, \ldots, e_m$ in the (finite dimensional) vector subspace of E spanned by the x_i, and a base $f_1, f_2, \ldots, f_n$ in the vector subspace of F spanned by the y_i. Then w can clearly be expressed in the form

$$w = \sum_{1 \leqslant r \leqslant m} \sum_{1 \leqslant s \leqslant n} \lambda_{rs} e_r \otimes f_s.$$

Since $w \neq o$, at least one λ_{rs} is different from zero; suppose that $\lambda_{11} \neq 0$. There is an element $x' \in E'$ with $\langle e_1, x' \rangle = 1$ but $\langle e_r, x' \rangle = 0$ for $2 \leqslant r \leqslant m$ (Ch. II, Lemma 5), and there is an element $y' \in F'$ with $\langle f_1, y' \rangle = 1$ but $\langle f_s, y' \rangle = 0$ for $2 \leqslant s \leqslant n$. Then

$$\langle w, x' \otimes y' \rangle = \lambda_{11} \neq 0.$$

Hence $\langle w, w' \rangle = 0$ for all $w' \in E' \otimes F'$ implies $w = o$. Similarly $\langle w, w' \rangle = 0$ for all $w \in E \otimes F$ implies $w' = o$, and so $(E \otimes F, E' \otimes F')$ is a dual pair.

PROPOSITION 8. *The topological tensor product $E \otimes F$ is separated if and only if both E and F are.*

Proof. Suppose first that E and F are separated. If E' and F' are their duals, (E, E') and (F, F') are dual pairs and therefore, by Lemma 1, $(E \otimes F, E' \otimes F')$ is a dual pair. But every element of $E' \otimes F'$ gives a continuous bilinear form on $E \times F$ and so belongs to the dual of $E \otimes F$. Hence $E \otimes F$ is separated (see Ch. II, § 3).

On the other hand, if, say, E is not separated, it contains a point $x \neq o$ which lies in every neighbourhood. If y is any point of F other than the origin, and $U \times V$ is any neighbourhood in $E \times F$, there is some $\lambda > 0$ with $\lambda y \in V$ and so

$$x \otimes y = \phi(\lambda^{-1} x, \lambda y) \in \phi(U, V).$$

But $x \otimes y \neq o$, and so $E \otimes F$ is not separated.

Even if E and F are complete, it is not in general true that $E \otimes F$ is complete. In applications it is often the completion of the topological tensor product that arises (see Suppl. 2); we denote this completion by $E \hat{\otimes} F$. When E and F are both metrisable, the elements of $E \hat{\otimes} F$ can be written as sums of convergent series of the form $\sum_{n=1}^{\infty} x_n \otimes y_n$. We deduce this from a more general result characterising the compact sets in $E \hat{\otimes} F$. First we require a lemma, which is of independent interest.

LEMMA 2. *Let E be a metrisable convex space, F a dense vector subspace and A a precompact subset of E. Then there is a sequence (x_n) of points of F, convergent to the origin, such that every element of A can be written in the form*

$$\sum_{n=1}^{\infty} \lambda_n x_n \quad \textit{with} \quad \sum_{n=1}^{\infty} |\lambda_n| \leqslant 1.$$

Proof. Let (U_n) be a base of absolutely convex neighbourhoods in E with $U_{n+1} \subseteq U_n$ for each n. Since A is precompact and F dense, there is a finite set B_1 of points of $F \cap (A + \tfrac{1}{2} U_1)$ with

$A \subseteq B_1 + \frac{1}{2}U_1$. Let $A_1 = (A - B_1) \cap \frac{1}{2}U_1$; then every point of A is of the form $\frac{1}{2}y_1 + z_1$ with $\frac{1}{2}y_1 \in B_1$ and $z_1 \in A_1$. Also A_1 is precompact and so there is a finite set $B_2 \subseteq F \cap (A_1 + \frac{1}{4}U_2)$ with $A_1 \subseteq B_2 + \frac{1}{4}U_2$. If $A_2 = (A_1 - B_2) \cap \frac{1}{4}U_2$, every point of A is now of the form $\frac{1}{2}y_1 + \frac{1}{4}y_2 + z_2$, with $\frac{1}{2}y_1 \in B_1, \frac{1}{4}y_2 \in B_2$ and $z_2 \in A_2$. Continuing this process we define finite subsets

$$B_n \subseteq F \cap (A_{n-1} + 2^{-n}U_n)$$

and precompact sets $A_n \subseteq 2^{-n}U_n$ with $A_n \subseteq B_{n+1} + 2^{-(n+1)}U_{n+1}$, such that every point x of A is of the form $\sum_{1 \leqslant r \leqslant n} 2^{-r}y_r + z_n$, where $2^{-r}y_r \in B_r$ for $1 \leqslant r \leqslant n$ and $z_n \in A_n$. Now $A_n \subseteq 2^{-n}U_n$ and so $z_n \to o$; thus x is of the form $\sum_{r=1}^{\infty} 2^{-r}y_r$. If now we take for (x_n) the points of $2B_1$, then those of $4B_2$, and so on, every point of A will have the required form and we shall have $x_n \to o$, since

$$2^rB_r \subseteq 2^r(A_{r-1} + 2^{-r}U_r) \subseteq 2U_{r-1} + U_r \subseteq 3U_{r-1}.$$

COROLLARY. *In a metrisable convex space any precompact set is contained in the closed absolutely convex envelope of a sequence convergent to the origin.*

For with the notation of Lemma 2, if

$$x = \sum_{n=1}^{\infty} \lambda_n x_n$$

and U is any neighbourhood, there is an n with

$$x - \sum_{1 \leqslant r \leqslant n} \lambda_r x_r \in U.$$

PROPOSITION 9. *Let E and F be metrisable convex spaces. Then $E \hat{\otimes} F$ is a Fréchet space and to each compact subset A of $E \hat{\otimes} F$ correspond sequences (x_n) in E and (y_n) in F, both convergent to the origin, such that every element of A can be written in the form*

$$\sum_{n=1}^{\infty} \lambda_n x_n \otimes y_n \quad \textit{with} \quad \sum_{n=1}^{\infty} |\lambda_n| \leqslant 1.$$

Proof. Let (U_n) and (V_n) be neighbourhood bases for E and F each consisting of a decreasing sequence of absolutely convex sets. The absolutely convex envelopes W_n of $\phi(U_n, V_n)$ form a neighbourhood base for $E \otimes F$, which is therefore metrisable. Hence $E \hat{\otimes} F$ is a Fréchet space. By Lemma 2, there is a sequence (w_m) of points of $E \otimes F$ with $w_m \to o$, such that every point $w \in A$ is of the form

$$\sum_{m=1}^{\infty} \mu_m w_m \quad \text{with} \quad \sum_{m=1}^{\infty} |\mu_m| \leqslant 1.$$

Since $w_m \to o$, there is a sequence $r(m) \to \infty$ with $w_m \in W_{r(m)}$ for each m. Then each w_m can be written as a finite sum

$$w_m = \sum_i \nu_{mi} x_{mi} \otimes y_{mi} \quad \text{with} \quad \sum_i |\nu_{mi}| \leqslant 1,\ x_{mi} \in U_{r(m)},\ y_{mi} \in V_{r(m)}$$

and so
$$w = \sum_{m=1}^{\infty} \sum_i (\mu_m \nu_{mi})\, x_{mi} \otimes y_{mi}.$$

If now the x_{mi} are relabelled (using dictionary order of suffixes) to form a sequence (x_n) convergent to o, the y_{mi} to form a sequence (y_n) convergent to o and the $\mu_m \nu_{mi}$ a sequence (λ_n) with

$$\sum_{n=1}^{\infty} |\lambda_n| \leqslant 1,$$

w will be expressed in the required form.

Corollary 1. *If E and F are metrisable convex spaces, every point of $E \hat{\otimes} F$ is of the form*

$$\sum_{n=1}^{\infty} \lambda_n x_n \otimes y_n \quad \textit{with} \quad \sum_{n=1}^{\infty} |\lambda_n| \leqslant 1,$$

$x_n \to o,\ y_n \to o$.

Corollary 2. *If E and F are metrisable convex spaces, every compact subset of $E \hat{\otimes} F$ is contained in the closed absolutely convex envelope of a sequence $(x_n \otimes y_n)$ with $x_n \to o$ and $y_n \to o$.*

It is a somewhat surprising fact that the topological tensor product of two barrelled metrisable spaces is itself barrelled. The proof, given below, has been divided up so that a preparatory lemma can be used to establish another special property of tensor products of barrelled metrisable spaces.

We recall that a bilinear mapping h of $E \times F$ into G defines two sets of partial linear mappings h_y of E into G for each fixed $y \in F$

and h_x of F into G for each fixed $x \in E$. If all these partial linear mappings are continuous, we say that h is *separately continuous* (or continuous in the variables separately). This is a consequence of, but does not necessarily imply, continuity of h as a bilinear mapping.

LEMMA 3. *Let E be a barrelled metrisable space and F a metrisable convex space, and let h be a separately continuous bilinear mapping of $E \times F$ into a third convex space G. Suppose either that F is also barrelled and B a barrel in G, or that B is a neighbourhood in G; then $h^{-1}(B)$ is a neighbourhood in $E \times F$.*

Proof. It is clearly sufficient, in the second case, to prove the result when B is closed and absolutely convex. Now in both cases, the continuity of h_a, for each $a \in E$, ensures that $V_a = h_a^{-1}(B)$ is (a barrel and therefore) a neighbourhood in F. If $h^{-1}(B)$ is not a neighbourhood in $E \times F$, there are sequences $x_n \to o$ in E and $y_n \to o$ in F with $h(x_n, y_n) \notin B$. Put

$$U = \bigcap_{n=1}^{\infty} h_{y_n}^{-1}(B).$$

Then U is absolutely convex and closed; it is also absorbent. For if $a \in E$, there is an n_a with $y_n \in V_a$ for $n \geqslant n_a$ and there is a scalar λ ($\geqslant 1$) with $h(a, y_n) \in \lambda B$ for $n < n_a$. Hence $h(a, y_n) \in \lambda B$ for all n and so U absorbs a. Thus U is a barrel in E and therefore a neighbourhood. Hence there is an n_1 with $x_n \in U$ for $n \geqslant n_1$ and so, for $n \geqslant n_1$, $h(x_n, y_n) \in B$. Thus $h^{-1}(B)$ must be a neighbourhood.

PROPOSITION 10. *The topological tensor product of two barrelled metrisable spaces is barrelled.*

Proof. Apply Lemma 3 to the canonical bilinear mapping ϕ of $E \times F$ into $E \otimes F$; if B is a barrel in $E \otimes F$, there is a neighbourhood $U \times V$ with $\phi(U, V) \subseteq B$.

PROPOSITION 11. *If E and F are metrisable convex spaces and one of them is barrelled, every separately continuous bilinear mapping defined on $E \times F$ is continuous on $E \times F$.*

Proof. Apply Lemma 3, taking for B any neighbourhood in G.

We end this section by mentioning briefly one other useful topology on a tensor product, and then taking a quick look at tensor products of normed spaces. Suppose that E and F are separated convex spaces with duals E' and F' and neighbourhood bases $\mathscr{U}$ and $\mathscr{V}$; let ϕ and ϕ' be the canonical mappings of $E \times F$ into $E \otimes F$ and of $E' \times F'$ into $E' \otimes F'$ respectively. Then $(E \otimes F, E' \otimes F')$ is a dual pair (Lemma 1) and the sets $\phi'(U^0, V^0)$ $(U \in \mathscr{U}, V \in \mathscr{V})$ satisfy the conditions of Chapter III, § 2, for defining on $E \otimes F$ a polar topology, which we shall call the topology of *equicontinuous convergence* on $E \otimes F$. The sets $(\phi'(U^0, V^0))^0$ form a base of neighbourhoods for this topology. Since $\phi(U, V) \subseteq (\phi'(U^0, V^0))^0$, the topology of equicontinuous convergence on $E \otimes F$ is coarser than the projective topology. We shall denote the completion of $E \otimes F$ under the topology of equicontinuous convergence by $E \check{\otimes} F$.

Finally, consider the situation when E and F are normed spaces with closed unit balls U and V. The closure of the absolutely convex envelope W of $\phi(U, V)$ can be taken as the closed unit ball in $E \otimes F$ under the projective topology; the norm turns out to be

$$\|w\|_p = \inf\{\sum_i \|x_i\| \cdot \|y_i\| : w = \sum_i x_i \otimes y_i\}.$$

For the topology of equicontinuous convergence, the norm is

$$\|w\|_e = \sup\{|\langle w, x' \otimes y' \rangle| : x' \in U^0, \ y' \in V^0\}.$$

Now $\|w\|_e \leqslant \|w\|_p$ for all $w \in E \otimes F$, since

$$|\langle \sum_i x_i \otimes y_i, x' \otimes y' \rangle| \leqslant \sum_i |\langle x_i, x' \rangle \langle y_i, y' \rangle| \leqslant \sum_i \|x_i\| \cdot \|y_i\|$$

for $x' \in U^0$, $y' \in V^0$. Also if w is of the form $x \otimes y$,

$$\begin{aligned} \|x\| \cdot \|y\| &= \sup\{|\langle x, x' \rangle \langle y, y' \rangle| : x' \in U^0, \ y' \in V^0\} \\ &= \|w\|_e \leqslant \|w\|_p \leqslant \|x\| \cdot \|y\|. \end{aligned}$$

Hence $\|x \otimes y\|_e = \|x \otimes y\|_p = \|x\| \cdot \|y\|$.

Consider now the dual of the normed space $E \otimes F$ (this is also the dual of $E \hat{\otimes} F$: Ch. VI, Prop. 6, Cor. 2). It is a Banach space under the norm

$$\begin{aligned} \|f\| &= \sup\{|f(w)| : w \in W\} = \sup\{|f(w)| : w \in \phi(U, V)\} \\ &= \sup\{|(f \circ \phi)(x, y)| : x \in U, y \in V\}. \end{aligned}$$

If $w' = \sum_i x'_i \otimes y'_i \in E' \otimes F'$, then w' can be regarded as a linear form on $E \otimes F$ and then

$$\|w'\| = \sup\{|\sum_i \langle x, x'_i\rangle \langle y, y'_i\rangle| : x \in U, y \in V\}$$

$$= \sup\{|\sum_i \langle x'', x'_i\rangle \langle y'', y'_i\rangle| : x'' \in U^{00}, y'' \in V^{00}\},$$

where U^{00} is the unit ball in the bidual E'' of E and V^{00} the unit ball in F'', since U and V are weakly dense in U^{00} and V^{00}. Thus the topology on $E' \otimes F'$ induced by this norm is that of equicontinuous convergence; $E' \check{\otimes} F'$ is a closed vector subspace of the dual of $E \hat{\otimes} F$ and $\|x' \otimes y'\| = \|x'\| . \|y'\|$.

3. The Krein–Milman theorem. Like the Hahn–Banach theorem, this result has a geometrical character. It is concerned with the problem of representing a given closed convex set A as the closed convex envelope of a small and conveniently identifiable subset K of itself, rather in the way in which, in a finite dimensional space, a convex polyhedron is determined by its vertices and a ball by its frontier. The theorem asserts that, in a convex space, if A is compact, K may be taken to be the set of its extremal points (as defined below). The main burden of the proof consists in showing that such a set A does in fact possess at least one extremal point.

In a vector space, a *line segment* $[a, b]$ is the set of points of the form $\lambda a + (1-\lambda) b$ with $0 \leqslant \lambda \leqslant 1$; the *extremities* of the line segment $[a, b]$ are the points a and b, the *interior* points those with $0 < \lambda < 1$. An *extremal point* of a subset A is a point of A which is an extremity of every line segment through it contained in A.

THEOREM 1. (Krein–Milman theorem.) *If E is a separated convex space and A is a convex compact subset of E, then A is the closed convex envelope of the set of its extremal points.*

Proof. It is sufficient to prove the theorem when E is a real space. For if E is a complex space, it may be regarded as a real space and the set A remains convex and compact; the set of extremal points of A is unchanged and so is its closed convex envelope.

Suppose now that E is a real space. Let B be the closed convex envelope of the set of extremal points of A (that extremal points always exist will appear in the course of the proof). Clearly $B \subseteq A$. If $B \neq A$, there is some point $z \in A$ with $z \notin B$. Since B is closed, there is an open convex set containing z and not meeting B. Hence (Ch. II, Prop. 5, Cor. 3) there is a continuous linear form f and a real number α with $f(x) \leqslant \alpha$ for all $x \in B$ and $f(z) > \alpha$. Since A is compact and f continuous, $\beta = \sup f(A)$ is finite; since $z \in A$, $\beta > \alpha$ and so $H = f^{-1}(\beta)$ does not meet B. If now we show that there is an extremal point of A belonging to H, so that H meets B, we shall have reached a contradiction, and thus we must have $B = A$.

Now the hyperplane H has the following property: if $[a, b]$ is a line segment in A with some interior point in H, then the whole segment lies in H. (This results from the definition of β and the convexity of H.) Generally, we call any translate of a closed vector subspace of E meeting A and with this property a *support* of A. Consider the set of supports of A contained in H. In this set, the (trivial) chain $\{H\}$ is contained in a maximal chain $\mathscr{M}$; let S be the intersection of the sets of $\mathscr{M}$. Then S is a support of A. For S is a translate of a closed vector subspace of E, S meets A because the intersections with A of the sets of $\mathscr{M}$ form a chain in a compact set (Ch. III, Prop. 11, Corollary) and has the required property because each set of $\mathscr{M}$ has. We show that S consists of one extremal point alone. Suppose that there are two points c and d in S; then there is a continuous linear form g with $g(c) \neq g(d)$ since E is separated. Now $S \cap A$ is compact; let $\gamma = \sup g(S \cap A)$ and let $G = S \cap g^{-1}(\gamma)$. We show that G is a support of A. First, G is a translation of a closed vector subspace of E. Next, G meets A because g, being continuous, attains the value γ on the compact set $S \cap A$. Finally, let $[a, b]$ be a line segment in A with an interior point $z \in G$. Then $z \in S$ and, S being a support, $[a, b]$ lies in S and so in $S \cap A$. Hence $g(x) \leqslant \gamma$ for all $x \in [a, b]$. But $g(z) = \gamma$ and z is an interior point and so $g(x) = \gamma$ for all $x \in [a, b]$. Thus $[a, b]$ lies in G. Hence G is a support of A, $G \subseteq S$, but $G \neq S$ since c and d cannot both belong to G. But this contradicts the maximality of $\mathscr{M}$. Hence S contains one point c alone. Finally, c is an extremal point. For if there is a line segment $[a, b]$ in A and c is an interior point, then by the support property of $S = \{c\}$, the whole segment $[a, b]$ lies in $\{c\}$, which is impossible. Thus we have shown that there is an extremal point of A belonging to H, and the whole theorem is proved.

SUPPLEMENT

(1) *Strict inductive limits.* The space $\mathscr{K}(]-\infty, \infty[)$ of continuous functions of compact support, under its inductive limit topology (Ch. I, Suppl. 2*c*), is an example of a strict inductive limit of a sequence of Fréchet spaces; the space $\mathscr{D}$ (Ch. I, Suppl. 3*d*) is another. In the case of $\mathscr{D}$, the defining subspaces $\mathscr{D}_n$ are Montel spaces (see Ch. IV, Suppl. 2) and it follows that $\mathscr{D}$ itself is a Montel space. More generally, any strict inductive limit E of a sequence of Montel spaces E_n is a Montel space. For every closed bounded set in E is contained in one of the E_n (Prop. 4) and so is compact; also E is barrelled (Ch. V, Prop. 6).

(2) *Tensor products.* If E is a given convex space and F is a space of real or complex valued functions on a set S, the space $E \otimes F$ can be identified with a space of functions on S taking their values in finite-dimensional vector subspaces of E, the typical element being a finite sum of the form $\sum_i x_i \phi_i$ with $x_i \in E$ and $\phi_i \in F$. For example, if F is the space $\mathscr{C}(S)$ of continuous functions on the compact space S, $E \otimes \mathscr{C}(S)$ is the space of continuous functions on S taking values in finite-dimensional subspaces of E. If we denote by $\mathscr{C}(S, E)$ the space of all continuous functions from S to E with the topology of uniform convergence, it is not difficult to show that $\mathscr{C}(S, E)$ induces on $E \otimes \mathscr{C}(S)$ the topology of equicontinuous convergence and that $E \otimes \mathscr{C}(S)$ is dense in $\mathscr{C}(S, E)$. Hence if E is complete (this in order that $\mathscr{C}(S, E)$ shall be complete), $E \check{\otimes} \mathscr{C}(S) = \mathscr{C}(S, E)$.

Taking for E the space $\mathscr{C}(T)$, where T is another compact space, and using the easily verified isomorphism between $\mathscr{C}(S, \mathscr{C}(T))$ and $\mathscr{C}(S \times T)$, we obtain the result

$$\mathscr{C}(S \times T) = \mathscr{C}(S) \check{\otimes} \mathscr{C}(T).$$

The theory of Lebesgue integration can be extended to functions taking their values in a Banach space E (see, for example,

Bourbaki, *Éléments de mathématique*. Livre VI. *Intégration*). Most of the classical results remain valid, with the norm replacing the absolute value in appropriate places. If $\mathscr{L}^1(S, E)$ denotes the space of integrable functions on the measure space S, taking values in E, then it can be shown that $E \otimes \mathscr{L}^1(S) = \mathscr{L}^1(S, E)$ (Grothendieck, *Produits tensoriels topologiques et espaces nucléaires*, Ch. I, § 2, Th. 2). In particular, $\mathscr{L}^1(S \times T) = \mathscr{L}^1(S) \hat{\otimes} \mathscr{L}^1(T)$.

In the above two examples it is essential to form the completion of the tensor product under the correct topology. But there is a large class of convex spaces F for which $E \hat{\otimes} F = E \check{\otimes} F$ for every convex space E. Such a convex space F is called *nuclear*. An infinite dimensional Banach space cannot be nuclear. On the other hand, the spaces $\mathscr{D}$, $\mathscr{E}$, $\mathscr{S}$ and $\mathscr{H}(D)$, together with their duals, are all nuclear. For nuclear spaces, their completed topological tensor products are often easily identified; for example, if E is complete, $E \hat{\otimes} \mathscr{H}(D)$ is the space of holomorphic functions from D to E.

Topological tensor products and nuclear spaces have been extensively studied by A. Grothendieck (*Produits tensoriels topologiques et espaces nucléaires* (1955)).

CHAPTER VIII

COMPACT LINEAR MAPPINGS

1. The Riesz theory. One of the early successes of functional analysis was an abstract theory of functional equations that included much of the Fredholm theory of integral equations. This theory, developed by F. Riesz, applies to linear integral equations of the form

$$\lambda x(u) - \int_a^b k(u, v)\, x(v)\, dv = y(u),$$

where k and y are given continuous functions, λ is a given scalar and x is the unknown (continuous) function. This equation is of the form $w_\lambda(x) = (\lambda i - t)(x) = y$, where i denotes the identity mapping of the space of functions considered and t is a linear mapping of this space into itself. In this form the problem of solving the equation is that of inverting the mapping w_λ. For this to be possible it is clearly necessary that $(\lambda i - t)$ be $(1, 1)$; in other words, that the only solution of the equation $t(x) = \lambda x$ should be zero. If the equation $t(x) = \lambda x$ has a non-zero solution x, λ is called an *eigenvalue* of the mapping t and the solution x is called a corresponding *eigenvector*.

In a finite-dimensional vector space, where linear mappings can be studied by means of matrices, the eigenvalues of t are the roots of the equation in λ: $\det(\lambda I - T) = 0$, where T is the matrix of t relative to some base, and w_λ has a (continuous) inverse if and only if λ is not an eigenvalue of t. In an infinite-dimensional convex space, the situation is not so simple; if λ is not an eigenvalue of the continuous mapping t, it still may happen that w_λ fails to map the space onto itself, and, even when w_λ has an inverse, it may not be continuous. The set of values of λ for which w_λ does not have a continuous inverse is called the *spectrum* of t. Thus an eigenvalue certainly belongs to the spectrum. The distinctive feature of the kind of mappings t that arise in the theory of integral equations is that, conversely, every point in the spectrum, with the possible exception of zero, is an eigenvalue; moreover, the spectrum consists of a sequence convergent to zero, together with zero, or a finite set. Now in a normed space the crucial property of t is that it (or sometimes some power t^k of it or even some polynomial in t) maps the unit ball into a compact set. A linear mapping that maps the unit ball into a compact set is compact, or completely continuous (for the formal definition in the general case see below). It was compact linear mappings that formed the subject of the elegant Riesz theory.

In this and the next section, we develop the theory of compact linear mappings in convex spaces. This section contains the Riesz theory generalised to convex spaces; in fact no essential use is made here of the hypothesis of local convexity. In the next section there is an account of the main duality properties of compact linear mappings. Here, of course, the theory depends on the assumption that the spaces are locally convex.

We begin the formal treatment by collecting together a few elementary properties of compact linear mappings. Let E and F be convex spaces over the complex field. The linear mapping t of E into F is called *compact* if there is some neighbourhood U in E and a compact set K in F with $t(U) \subseteq K$. Then t is continuous, for if V is any neighbourhood in F, $t(\lambda U) \subseteq \lambda K \subseteq V$ for all sufficiently small $\lambda > 0$. If s is a continuous linear mapping into E and u a continuous linear mapping defined on F then $s \circ t$ and $t \circ u$ are both compact. It follows that if t is a compact linear mapping of E into itself, then any polynomial in t without constant term is compact (since a finite sum of compact sets is compact).

In most of this section, we study the linear mapping $w_\lambda = \lambda i - t$, where t is compact and λ is not zero. When λ is fixed, we usually write w instead of w_λ. Much of the Riesz theory remains applicable if, instead of being given that the linear mapping itself is compact, we know only that some power of it (or some polynomial in it) is compact. We cope with this in the following way. Suppose, for example, that s^k is compact for some positive integer $k \geqslant 2$. Then $v = \mu i - s$ is a factor of $w = \mu^k i - s^k$; if the other factor is u we have

$$u \circ v = v \circ u = \lambda i - t,$$

where $\lambda = \mu^k \neq 0$ if $\mu \neq 0$ and $t = s^k$ is compact. We therefore study continuous linear mappings v that satisfy the identity

$$u \circ v = v \circ u = \lambda i - t,$$

where u is continuous, t is compact and $\lambda \neq 0$.

PROPOSITION 1. *Suppose that t is a compact linear mapping of the separated convex space E into itself, that $\lambda \neq 0$ and that u and v are two continuous linear mappings of E into itself such that*

$$u \circ v = \lambda i - t.$$

Then $v^{-1}(o)$ is finite-dimensional, v is an open mapping of E onto $v(E)$ and $v(E)$ is closed in E.

Proof. Since t is compact, there is a neighbourhood U mapped by t into a compact set K. Let $N = v^{-1}(o)$. Then if

$$x \in U \cap N, \quad u(v(x)) = o$$

and so $\lambda x = t(x)$. Hence $\lambda x \in t(U) \subseteq K$ and so $U \cap N \subseteq \lambda^{-1}K$. Thus N has a precompact neighbourhood and is therefore finite-dimensional (Ch. III, Th. 2).

Next let $\mathscr{V}$ be a base of balanced neighbourhoods for E. If v is not open there is some $W \in \mathscr{V}$, which we may clearly suppose contained in U, such that $v(W)$ is not a neighbourhood in $v(E)$. Then each $V \in \mathscr{V}$ meets $v(E) \cap \sim v(W) = v(\sim (W+N))$; if x is a common point, there is a μ with

$$0 < \mu \leqslant 1 \quad \text{and} \quad \mu x \in v(2W) \cap \sim v(W).$$

Hence V also meets $v(A)$, where $A = 2W \cap \sim (W+N)$. Thus the sets $A \cap v^{-1}(V)$ form the base of a filter $\mathscr{F}$ to which A belongs, and $v(\mathscr{F}) \to o$. Now $t(\mathscr{F})$ contains $t(A) \subseteq t(2W) \subseteq 2K$; hence $\mathscr{F}$ has a refinement $\mathscr{G}$ with $t(\mathscr{G})$ convergent to some point $z \in 2K$. Since $\lambda i = t + u \circ v$, $\lambda \mathscr{G} \to z + u(o) = z$. But $\lambda A \in \lambda \mathscr{G}$ and so $z \in \lambda \bar{A}$, while also $v(\lambda \mathscr{G}) = \lambda v(\mathscr{G}) \to o$ so that $v(z) = o$. Thus $z \in N$ and so $\lambda^{-1} z \in N \cap \bar{A}$. But $N + W$ does not meet A, and this contradiction proves that v is open.

Finally, $v(E)$ is closed. For, if $a \in \overline{v(E)}$, the sets $v(E) \cap (a + V)$ with $V \in \mathscr{V}$ form the base of a Cauchy filter on $v(E)$. Hence there is a set small of order $v(U)$; if $v(b)$ is any point in this set, the sets $(b+U) \cap v^{-1}(a+V)$ with $V \in \mathscr{V}$ form the base of a filter $\mathscr{F}$ with $v(\mathscr{F}) \to a$. Now $t(\mathscr{F})$ contains $t(b+U) \subseteq t(b) + K$, a compact set, and so $\mathscr{F}$ has a refinement $\mathscr{G}$ with $t(\mathscr{G})$ convergent to some point y, say. Since $\lambda i = t + u \circ v$, $\lambda\mathscr{G} \to y + u(a)$ and so

$$v(\lambda\mathscr{G}) \to v(y+u(a)) \in v(E).$$

But also $v(\lambda\mathscr{G}) = \lambda v(\mathscr{G}) \to \lambda a$, so that $a \in v(E)$ and $v(E)$ is closed.

COROLLARY. *Suppose also that u and v commute (i.e. $u \circ v = v \circ u$). Then for any integer $r \geqslant 1$, $v^{-r}(o)$ is finite dimensional and v^r is an open mapping of E onto the closed vector subspace $v^r(E)$.* (By $v^{-r}(o)$ we mean $(v^r)^{-1}(o)$, the set of x with $v^r(x) = o$.)

For then $u^r \circ v^r = (u \circ v)^r = (\lambda i - t)^r = \lambda^r i - s$, where s is a polynomial in t without constant term and is therefore compact.

Of course, when u and v commute, it is equally true that $u^{-r}(o)$ is finite dimensional and that u^r is an open mapping of E onto the closed vector subspace $u^r(E)$.

We require next some algebraic properties of the vector subspaces $v^{-r}(o)$ and $v^r(E)$ which are independent of the special form that v has here.

If v is any linear mapping of a vector space E into itself, the vector subspaces $\{o\} = v^{-0}(o), v^{-1}(o), v^{-2}(o), \ldots$, form an increasing sequence. If there is any integer $n \geqslant 0$ with $v^{-n}(o) = v^{-n-1}(o)$, then

$$v^{-n-2}(o) = v^{-1}(v^{-n-1}(o)) = v^{-1}(v^{-n}(o)) = v^{-n-1}(o) = v^{-n}(o)$$

and this argument can be continued to show that $v^{-r}(o) = v^{-n}(o)$ for all $r \geqslant n$. Hence either the sequence $(v^{-r}(o) : r \geqslant 0)$ is strictly increasing, or there is a least integer $n \geqslant 0$ such that $v^{-r}(o)$ are all distinct for $0 \leqslant r \leqslant n$ and subsequent subspaces are identical with $v^{-n}(o)$. In this latter case we say that v has *finite ascent* n.

Similarly the vector subspaces $E = v^0(E), v(E), v^2(E), \ldots$, form a sequence which either decreases strictly or has a least integer $m \geqslant 0$ such that $v^r(E)$ are all distinct for $0 \leqslant r \leqslant m$ and subsequent subspaces coincide with $v^m(E)$. In this latter case we say that v has *finite descent* m.

LEMMA 1. *For any linear mapping v of E into itself and any non-negative integers r, s,*

(i) $v^{-r}(o) = v^{-r-s}(o)$ *if and only if* $v^r(E) \cap v^{-s}(o) = \{o\}$,

(ii) $v^s(E) = v^{s+r}(E)$ *if and only if* $v^r(E) + v^{-s}(o) = E$.

Proof. Both sides of (i) assert that if $y = v^r(x)$ is mapped into the origin by v^s then $y = o$. Also both sides of (ii) assert that for all $x \in E$, there is some $y \in E$ with $x - v^r(y) \in v^{-s}(o)$.

LEMMA 2. *If the linear mapping v of E into itself has finite ascent n and finite descent m, then $m = n$ and E is the (algebraic) direct sum of $v^{-n}(o)$ and $v^n(E)$.*

Proof. By (i) of Lemma 1, with $r = n$ and $s = 1$, we have $v^n(E) \cap v^{-1}(o) = \{o\}$. If $m \leqslant n$ this is the same as

$$v^m(E) \cap v^{-1}(o) = \{o\}$$

and so, applying (i) again with $r = m$ and $s = 1$, $v^{-m}(o) = v^{-m-1}(o)$. But this implies $m = n$.

Similarly we can apply (ii) with $s = m$ and $r = 1$ to show that $v(E) + v^{-m}(o) = E$. If $m \geqslant n$ this gives $v(E) + v^{-n}(o) = E$ and therefore, by (ii) with $s = n$ and $r = 1$, $v^n(E) = v^{n+1}(E)$. Hence $m = n$.

Finally, the left sides of (i) and (ii) are both satisfied when $r = s = n$ and so the right sides are valid. This shows that E is the direct sum of $v^n(E)$ and $v^{-n}(o)$.

We now return to the study of compact linear mappings, showing that, if $u \circ v = v \circ u = \lambda i - t$, then v has finite ascent and descent. For this we require:

LEMMA 3. *Suppose that t is a linear mapping of E into itself and that U is a neighbourhood such that $t(U)$ is contained in a compact set. Suppose also that $w = \lambda i - t$ with $\lambda \neq 0$, and that F and G are distinct vector subspaces of E with G closed, $G \subseteq F$ and $w(F) \subseteq G$. Then there is a point $x \in F \cap (2U)$ with $x \notin G + U$.*

Proof. First, the conditions ensure that $F \nsubseteq G + U$. Otherwise, $F \subseteq G + \epsilon\lambda U \cap F \subseteq G + \epsilon w(F) + \epsilon t(U) \subseteq G + \epsilon t(U)$ for every $\epsilon > 0$, since $\lambda i = w + t$. Hence, given any neighbourhood V, $\epsilon t(U) \subseteq V$ for some $\epsilon > 0$ and so $F \subseteq G + V$. But G is closed and so $F = G$, contrary to hypothesis.

Next, $F \cap (2U) \nsubseteq G + U$, for otherwise we should have

$$F \cap 2^n U \subseteq G + (F \cap 2^{n-1}U) \subseteq G + G + (F \cap 2^{n-2}U) \subseteq \ldots \subseteq G + U,$$

and hence $F \subseteq G + U$, contrary to what has been proved. There is therefore a point x with the required properties.

PROPOSITION 2. *Suppose that t is a compact linear mapping of the separated convex space E into itself, that $\lambda \neq 0$ and that u and v are two commuting continuous linear mappings such that*

$$u \circ v = v \circ u = \lambda i - t.$$

Then u and v have finite ascent and finite descent.

Proof. There is a neighbourhood U mapped by t into a compact set K. Write

$$w = u \circ v = v \circ u = \lambda i - t.$$

Suppose first that v does not have finite ascent. Then the closed vector subspaces $N_r = v^{-r}(o)$ are all distinct. Now, if $x \in N_{r+1}$, then $v^{r+1}(x) = o$ and so $v^r(w(x)) = u(v^{r+1}(x)) = u(o) = o$ so that $w(x) \in N_r$. Thus $w(N_{r+1}) \subseteq N_r$ and Lemma 3 applies with $F = N_{r+1}$, $G = N_r$. There is, therefore, for each $r \geqslant 0$, a point $x_r \in N_{r+1} \cap (2U)$ with $x_r \notin N_r + U$. Then $t(x_r) \in 2K$ and, if $r > s$,

$$t(x_r) - t(x_s) = \lambda x_r - w(x_r) - \lambda x_s + w(x_s) \in \lambda x_r + N_r.$$

Hence $t(x_r) - t(x_s) \notin \lambda U$. But this contradicts the precompactness of $2K$ and so v must have finite ascent.

Next suppose that v does not have finite descent. Then the vector subspaces $M_r = v^r(E)$, which are closed, by Proposition 1, are all distinct. Again $w(M_r) \subseteq M_{r+1}$ and so Lemma 3 applies with $F = M_r$, $G = M_{r+1}$. As before, there are points $y_r \in M_r \cap (2U)$ with $y_r \notin M_{r+1} + U$. Then $t(y_r) \in 2K$ and for $r < s$

$$t(y_r) - t(y_s) \in \lambda y_r + M_{r+1}.$$

This leads to a similar contradiction and so v must have finite descent.

Corollary 1. *If n is the ascent of v, then E is the topological direct sum of $v^{-n}(o)$ and $v^n(E)$.*

Proof. By the proposition and Lemma 2, n is also the descent of v and E is the algebraic direct sum of $v^{-n}(o)$ and $v^n(E)$. But, by the Corollary to Proposition 1, $v^{-n}(o)$ is finite-dimensional and $v^n(E)$ is closed, and so (Ch. v, Prop. 29, Corollary) E is their topological direct sum.

Corollary 2. *If t is a compact linear mapping of the separated convex space E into itself and $w = \lambda i - t$, where $\lambda \neq 0$, the following are equivalent:*

(i) *λ is not an eigenvalue of t;*
(ii) *w is $(1, 1)$;*
(iii) *the ascent of w is zero;*
(iv) *the descent of w is zero;*
(v) *w maps E onto itself.*

(By the proposition, with $u = i$, $v = w$.)

THEOREM 1. *Let t be a compact linear mapping of the separated convex space E into itself and let u and v be two commuting continuous linear mappings such that $u \circ v = v \circ u = \lambda i - t$, where $\lambda \neq 0$. Then E can be written as the topological direct sum of two (closed) vector subspaces M and N, each mapped into itself by v. On M, v is an isomorphism, while N is finite dimensional and on it v is nilpotent* (i.e. *there is an n with $v^n = o$*). *For each positive integer r, $v^{-r}(o)$ and $E/v^r(E)$ have the same dimension. Finally, on E, v can be written in the form $v = v_1 + v_2$, where v_1 is an isomorphism of E onto itself and v_2 maps E into a finite-dimensional vector subspace.*

Proof. If n is the ascent of v, then (Prop. 2, Cor. 1) E is the topological direct sum of $M = v^n(E)$ and $N = v^{-n}(o)$. By the Corollary of Proposition 1, N is finite-dimensional; also it is clear that $v(N) \subseteq N$ and $v^n = o$ on N. Next, $v(M) = M$ and

$$u(M) = u(v^n(E)) = v^n(u(E)) \subseteq v^n(E) = M,$$

so that $u(v(M)) \subseteq M$ and therefore $t(M) \subseteq M$. Hence Proposition 1 can be applied to t, u and v restricted to M; v is open on M. But v is also (1, 1) on M and so v is an isomorphism on M. Also

$$E/v^r(E) = (M+N)/(v^r(M)+v^r(N)) = (M+N)/(M+v^r(N)),$$

which is (algebraically) isomorphic to $N/v^r(N)$. Since N is finite-dimensional this has the same dimension as $v^{-r}(o)$. To define v_1 and v_2 let p and q be the projections of E onto M and N; then $v_1 = v \circ p + q$ is an isomorphism of E onto itself and $v_2 = v \circ q - q$ maps E into N, while

$$v_1 + v_2 = v \circ p + v \circ q = v \circ (p+q) = v.$$

COROLLARY 1. *Suppose that s is a continuous linear mapping of the separated convex space E into itself and that there is a polynomial ϕ with $\phi(s)$ compact. Then, if $\phi(\mu) \neq 0$, the mapping $v = \mu i - s$ has all the properties enunciated for the mapping v in the theorem.*

Proof. Let $\psi(\xi)$ be the quotient when $\phi(\xi)$ is divided by $-(\mu - \xi)$ so that

$$\psi(\xi)(\mu - \xi) = \phi(\mu) - \phi(\xi).$$

Then with $u = \psi(s)$, $v = \mu i - s$, $\lambda = \phi(\mu) \neq 0$ and $t = \phi(s)$, the conditions of the theorem are satisfied and so the conclusions are valid.

COROLLARY 2. *If t is compact and $w = \lambda i - t$, where $\lambda \neq 0$, then w has all the properties enunciated for the mapping v in the theorem. In addition, on the subspace M on which w is an isomorphism, the inverse is of the form $w^{-1} = \lambda^{-2}(\lambda i - s)$, where s is a compact linear mapping of M into itself, commuting with t. In particular, if λ is not zero and not an eigenvalue of t, w^{-1} has the above form on the whole of E.*

Proof. Certainly w^{-1} can be written in the form $\lambda^{-2}(\lambda i - s)$ for some continuous linear mapping s of M into itself. Then on M,

$$i = w \circ w^{-1} = (\lambda i - t) \circ \lambda^{-2}(\lambda i - s) = i - \lambda^{-1}t - \lambda^{-1}s + \lambda^{-2}t \circ s$$

and also

$$i = w^{-1} \circ w = \lambda^{-2}(\lambda i - s) \circ (\lambda i - t) = i - \lambda^{-1}s - \lambda^{-1}t + \lambda^{-2}s \circ t.$$

Hence $s \circ t = t \circ s$ and $s = \lambda^{-1}t \circ (-\lambda i + s)$. Since t is compact and $-\lambda i + s$ continuous, s is compact.

There are various different ways of looking at the results stated in Theorem 1 and its Corollaries. In terms of the equation $w_\lambda(x) = \lambda x - t(x) = y$, it states that if t is compact and $\lambda \neq 0$ (or more generally, if some polynomial $\phi(t)$ is compact and $\phi(\lambda) \neq 0$) then either the equation has a unique solution $x = w^{-1}(y)$ for each y or there is at least one $x \neq o$ with $w(x) = o$ (the Fredholm alternative). In the latter case, the set of x with $w(x) = o$ is finite-dimensional.

Another conclusion is that every point λ of the spectrum of a compact linear mapping t, except possibly for $\lambda = 0$, is an eigenvalue with a finite-dimensional subspace of eigenvectors. (When E is infinite-dimensional, $\lambda = 0$ belongs to the spectrum of t; otherwise t^{-1} would be continuous and E would have a compact neighbourhood.) The last results of this section give more information about the spectrum.

LEMMA 4. *For any linear mapping t of a vector space E into itself, eigenvectors corresponding to distinct eigenvalues are linearly independent.*

Proof. Otherwise, there is an integer $n \geqslant 0$ such that any n eigenvectors (corresponding to distinct eigenvalues) are linearly independent, but that there are $(n+1)$ such eigenvectors

$x_0, x_1, \ldots, x_n$ and non-zero scalars $\mu_0, \mu_1, \ldots, \mu_n$ with

$$\sum_{0 \leqslant r \leqslant n} \mu_r x_r = o.$$

If, for each r, $t(x_r) = \lambda_r x_r$, this gives, on applying $\lambda_0 i - t$,

$$\sum_{1 \leqslant r \leqslant n} \mu_r(\lambda_0 - \lambda_r) x_r = o.$$

This shows that $x_1, x_2, \ldots, x_n$ are linearly dependent, which is a contradiction.

PROPOSITION 3. *In a separated convex space, the spectrum of a compact linear mapping consists of either a finite set or a sequence convergent to zero.*

Proof. We have already seen that, apart possibly from $\lambda = 0$, the points of the spectrum of the compact linear mapping t are all eigenvalues. Take $\epsilon > 0$ and suppose that there is a sequence (λ_r) of distinct eigenvalues with $|\lambda_r| \geqslant \epsilon$. Let (x_r) be a sequence of corresponding eigenvectors and H_n the vector subspace spanned by the first n of these. Then the H_n are closed, and all distinct, by Lemma 4. Also if $w_r = \lambda_r i - t$, $w_r(H_r) \subseteq H_{r-1}$. Hence, if U is a balanced neighbourhood mapped by t into a compact set K, Lemma 3 applies with $F = H_r$, $G = H_{r-1}$, $w = w_r$. There is therefore, for each $r > 1$, a point $y_r \in H_r \cap (2U)$ with $y_r \notin H_{r-1} + U$. Then $t(y_r) \in 2K$ and, if $r > s$,

$$t(y_r) - t(y_s) = \lambda_r y_r - w_r(y_r) - \lambda_s y_s + w_s(y_s) \in \lambda_r y_r + H_{r-1}.$$

Hence $t(y_r) - t(y_s) \notin \epsilon U$. This contradicts the precompactness of $t(U)$. There are therefore only finitely many eigenvalues λ with $|\lambda| \geqslant \epsilon$, and the result follows.

COROLLARY. *Suppose that s is a linear mapping of E into itself and that there is a polynomial ϕ (not a constant) with $\phi(s)$ compact. Then the spectrum of s is finite or countable and its only limit points are zeros of ϕ.*

Proof. If $s(x) = \lambda x$, then $\phi(s)(x) = \phi(\lambda) x$ and so the image by ϕ of the spectrum of s is finite or a sequence convergent to zero.

(In fact the corollary is still valid if ϕ is a constant other than zero; for then i is compact, E finite-dimensional and the spectrum of s a finite set.)

2. Duality theory. We continue with the study of compact linear mappings, dealing in this section with properties involving the dual space and the transpose mapping. In the first place, the transpose can always be made compact:

LEMMA 5. *If t is a compact linear mapping of the separated convex space E into itself, its transpose t' is compact when the dual E' of E has the topology of uniform convergence on the absolutely convex compact subsets of E.*

Proof. There is a neighbourhood U, which we may take absolutely convex, with $t(U)$ contained in a compact set. Then $V' = (t(U))^0$ is a neighbourhood in E'. Also (Ch. II, Lemma 6) $(t(U))^0 = t'^{-1}(U^0)$ so that $t'(V') \subseteq U^0$. Now it follows from Proposition 4 of Chapter VI that U^0 is compact in the stated topology on E', so that t' is compact.

PROPOSITION 4. *Let t be a compact linear mapping of the separated convex space E into itself, let λ be non-zero, and let u and v be two commuting continuous linear mappings such that*

$$u \circ v = v \circ u = \lambda i - t.$$

Also let t', u' and v' be the transposes of t, u and v, and i' the identity mapping of the dual E' of E onto itself, so that

$$u' \circ v' = v' \circ u' = \lambda i' - t'.$$

Then:

(i) *v and v' have the same (finite) ascent and descent n;*

(ii) *for each positive integer r, $v^{-r}(o)$ and $v'^{-r}(o)$ have the same dimension;*

(iii) *writing $M = v^n(E)$ and $N = v^{-n}(o)$, E' is the topological direct sum of their polars $M^0 = v'^{-n}(o)$ and $N^0 = v'^n(E')$ under any topology for which t' is compact (for example, the topology mentioned in Lemma 5). On the finite dimensional space M^0, v' is nilpotent, and on N^0, v' is an isomorphism.*

Proof. For any positive integer r, we have (Ch. II, Lemma 6) $(v^r(E))^0 = v'^{-r}(o)$ and $(v'^r(E'))^0 = v^{-r}(o)$. Since $v^r(E)$ is closed (Prop. 1) and $v'^r(E')$ is closed in E' under any topology of the dual pair (E', E) (Lemma 5 and Prop. 1), we also have $v^r(E) = (v'^{-r}(o))^0$ and $v'^r(E') = (v^{-r}(o))^0$. These relations show that the ascent of v'

is equal to the descent of v, and that the descent of v' is equal to the ascent of v, and thus furnish a proof of (i). For (ii), the dual of $v^{-r}(o)$ is $E'/(v^{-r}(o))^0 = E'/v'^r(E')$, which, by Theorem 1, has the same dimension as $v'^{-r}(o)$. Finally, (iii) is the result of applying Theorem 1 to t' under any suitable topology.

COROLLARY. *Apart possibly from $\lambda = 0$, t' and t have the same eigenvalues. If λ is a non-zero eigenvalue, the equation, in x, $\lambda x - t(x) = y$ has a solution if and only if y vanishes on $w'^{-1}(o)$ and the equation, in x', $\lambda x' - t(x') = y'$ has a solution if and only if y' vanishes on $w^{-1}(o)$.* (Here $w = \lambda i - t$ and w' is the transpose of w.)

(By the proposition with $u = i$ and $v = w$.)

In the statement of Proposition 4 (iii) we envisage the possibility of other topologies under which t' may be compact. This is in fact realised when, for example, E is a Banach space and E' has its norm topology $\beta(E', E)$; a theorem of Schauder proves that t is compact if and only if t' is compact. It is not always true in a convex space that the transpose of a compact linear mapping is compact in the strong topology on the dual. To obtain the natural analogue of Schauder's theorem for a convex space we have to consider linear mappings that map bounded sets into compact sets. (In a normed space, this notion clearly coincides with that of a compact linear mapping.)

LEMMA 6. *Suppose that E and F are separated convex spaces and that t is a weakly continuous linear mapping of E into F. Let E' have the topology of $\mathscr{A}$-convergence. Then t maps the sets of $\mathscr{A}$ into precompact sets if and only if t' maps equicontinuous sets into compact sets.*

Proof. By Theorem 3 of Chapter III, each $t(A)$ is precompact if and only if t' maps equicontinuous sets into precompact sets. Further, t' is weakly continuous and an equicontinuous set B' is a subset of a weakly compact set, and so $t'(B')$ is contained in a weakly compact set. Hence the weak closure of $t'(B')$ is weakly complete, and so complete in the topology of $\mathscr{A}$-convergence (Ch. VI, Prop. 3, Corollary). Thus, in the topology of $\mathscr{A}$-convergence, the closure of $t'(B')$ is complete and, being precompact, is compact; hence t' maps equicontinuous sets into compact sets.

Proposition 5. *Suppose that E and F are separated convex spaces, that their duals E' and F' have their strong topologies $\beta(E',E)$ and $\beta(F',F)$ and that t is a weakly continuous linear mapping of E into F. Suppose also that F is complete and barrelled. Then t maps bounded sets into compact sets if and only if t' has the same property.*

Proof. Suppose that t maps bounded sets into compact sets. Since F is a barrelled space, the $\beta(F',F)$-bounded sets are equicontinuous and so t' maps them into compact sets (Lemma 6). Conversely, if t' has this property then t maps bounded sets into precompact sets (Lemma 6) and so into compact sets since F is complete.

The conditions on F are satisfied in particular when F is a Fréchet space.

Theorem 2. (Schauder.) *Let E and F be Banach spaces, E' and F' their duals with the norm topologies and t a weakly continuous linear mapping of E into F. Then t is compact if and only if t' is compact.*

Proof. The unit balls in E and F' are each simultaneously neighbourhoods and bounded sets; thus the theorem follows from Proposition 5.

Finally, we add here the corresponding theorem about mappings in Banach spaces that take bounded sets into weakly compact sets. Again we make use of the completion theory of Chapter VI, § 1, in a preparatory lemma:

Lemma 7. *Suppose that E and F are separated convex spaces with duals E' and F' and that E' has a topology of $\mathscr{A}$-convergence, under which its dual is G. Suppose that t is a weakly continuous linear mapping of E into F, that t' is the transpose of t and t'' the transpose of t'. Then*

(i) *t maps the sets of $\mathscr{A}$ into absolutely convex $\sigma(F,F')$-compact sets if and only if $t''(G) \subseteq F$;*

(ii) *t' maps the equicontinuous sets into $\sigma(E',G)$-compact sets if and only if $t''(G) \subseteq \hat{F}$.*

Proof. (i) Since $G = \bigcup_{A \in \mathscr{A}} A^{00}$,

$$t''(G) = \bigcup t''(A^{00}) \subseteq \bigcup (t(A))^{00},$$

by applying Lemma 6 of Chapter II twice. Now if t maps $A \in \mathscr{A}$ into an absolutely convex $\sigma(F, F')$-compact set, then $(t(A))^{00}$, being the $\sigma(F'^*, F')$-closure of the absolutely convex envelope of $t(A)$, lies in this compact set and so in F. Thus $t''(G) \subseteq F$. Conversely, if $t''(G) \subseteq F$, then, since t'' is continuous under the topologies $\sigma(G, E')$ and $\sigma(F'^*, F')$ and A^{00} is $\sigma(G, E')$-compact, $t''(A^{00})$ is $\sigma(F, F')$-compact and $t(A) \subseteq t''(A^{00})$. (ii) If $t''(G) \subseteq \hat{F}$ then t' is continuous under the topologies $\sigma(F', \hat{F})$ and $\sigma(E', G)$. Now on an equicontinuous set the topologies $\sigma(F', \hat{F})$ and $\sigma(F', F)$ coincide (Ch. VI, Th. 2, Cor. 3), and, the closure of an equicontinuous set being $\sigma(F', F)$-compact, its image is $\sigma(E', G)$-compact. Conversely, let $\mathscr{V}$ be a base of neighbourhoods in F and let $z \in G$. Since

$$\hat{F} = \bigcap_{V \in \mathscr{V}} (F + V^{00})$$

(Ch. VI, Th. 1), it is sufficient to show that for each $V \in \mathscr{V}$, $t''(z) \in F + V^{00}$. Now $z \in A^{00}$ for some $A \in \mathscr{A}$, and A^{00}, being the $\sigma(G, E')$-closure of the absolutely convex envelope of A, is also its $\tau(G, E')$-closure. But if $V \in \mathscr{V}$, $t'(V^0)$ is absolutely convex and contained in a $\sigma(E', G)$-compact set, and so its polar is a $\tau(G, E')$-neighbourhood. Hence there is some $x \in E$ with

$$z \in x + (t'(V^0))^0 = x + t''^{-1}(V^{00}).$$

Thus $t''(z) \in F + V^{00}$.

THEOREM 3. *Suppose that E is a Banach space, with dual E' and bidual E'', that F is a Banach space with dual F', and that t is a weakly continuous linear mapping of E into F, with transpose t' and bitranspose t''. Then the following are equivalent:*

(i) *t maps bounded sets into $\sigma(F, F')$-compact sets;*

(ii) *t' maps bounded sets into $\sigma(E', E'')$-compact sets;*

(iii) *$t''(E'') \subseteq F$.*

Proof. This follows from Lemma 7 since F is complete and bounded subsets of F' are equicontinuous.

APPENDIX

Since this book was written, various new closed graph theorems have been discovered. Compared with those described in Chapter VI, the conditions on the range space are relaxed at the expense of those on the domain space, but both sets of conditions are verified in practice by a wide variety of useful spaces. The structures needed for the range space are of a countable type not previously studied in this tract; they were introduced by M. De Wilde ('Sur le théorème du graphe fermé', *C.R. Acad. Sci. Paris, Ser. A–B*, 265 (1967) 376–9; *Réseaux dans les espaces linéaires à semi-normes*, Mém. Soc. Roy. Sci. Liège, XVIII (1969) Fasc. 2). In this Appendix we give a brief account of this theory.

1. Spaces with webs. A *web* $\mathscr{W}$ in a vector space F is a countable family of absolutely convex subsets, indexed by the finite sequences of positive integers and arranged in *layers*. The first layer consists of a sequence (A_i) whose union absorbs each point of F. For each set A_i, there is a sequence (A_{ij}) of subsets of $\frac{1}{2}A_i$, whose union, as j varies, absorbs each point of A_i. For convenience later, we shall call the sets A_{ij} the sets *determined* by A_i. All these sets A_{ij}, as i and j both vary, constitute the second layer of $\mathscr{W}$. Then for each set A_{ij} there is a sequence (A_{ijk}) of subsets of $\frac{1}{2}A_{ij}$, whose union, as k varies, absorbs each point of A_{ij}. These are the sets determined by A_{ij}, and the sets A_{ijk} as i, j and k vary constitute the third layer of $\mathscr{W}$. And so on.

By a *strand* of the web $\mathscr{W}$ we shall mean any sequence of sets A_i, A_{ij}, A_{ijk}, ..., one from each layer, each after the first being one of the sets determined by its predecessor. Thus each infinite sequence (i_n) of positive integers determines a strand whose general term is A_ϕ, where $\phi = (i_1, i_2, ..., i_n)$. Writing out the index in full for each set of a strand leads to a cumbersome notation, but this can be avoided wherever it is sufficient to consider one strand at a time. We can then denote a typical strand by (W_n). It is a consequence of the above definition that, for all n, $W_{n+1} \subseteq \frac{1}{2} W_n$ for every strand (W_n).

In the definition of a web, we required inclusions of the form $A_{ij} \subseteq \frac{1}{2} A_i$, $A_{ijk} \subseteq \frac{1}{2} A_{ij}$, ... A family $\mathscr{W}'$ of subsets satisfying all the other conditions for a web can be made into a web $\mathscr{W}$ with sets of the form

$$A_i = A_i', \quad A_{ij} = \tfrac{1}{2}(A_i' \cap A_{ij}'),$$
$$A_{ijk} = \tfrac{1}{4}(A_i' \cap A_{ij}' \cap A_{ijk}'), \quad \ldots.$$

Suppose now that F is a separated convex space. We shall say that a web $\mathscr{W}$ on F is *compatible* with the topology if, for each strand (W_n) of $\mathscr{W}$ and each neighbourhood V, there is an n with $W_n \subseteq V$. Thus $\mathscr{W}$ is compatible if every strand contains arbitrarily small sets, or equivalently, for each strand (W_n), $x_n \in W_n$ implies $x_n \to o$. Indeed, when $\mathscr{W}$ is compatible and, for some strand, $x_n \in W_n$ for all n, the series Σx_n has Cauchy partial sums. For if V is any neighbourhood and $m > n$,

$$s_m - s_n = x_{n+1} + \ldots + x_m \in W_{n+1} + \ldots + W_m$$
$$\subseteq \tfrac{1}{2} W_n + \ldots + \frac{1}{2^{m-n}} W_n \subseteq W_n \subseteq V$$

for sufficiently large m, n.

In a separated convex space F, we shall call a web $\mathscr{W}$ a *completing* web if, for each strand (W_n), the series Σx_n is convergent for every choice of $x_n \in W_n$. A completing web is certainly compatible, and the converse holds in any space in which every Cauchy sequence is convergent (we might call such a space *sequentially complete*). We note this fact for reference later.

Lemma 1. *In a (sequentially) complete space, a compatible web is completing.*

Many important classes of convex spaces have completing webs. We give two immediately; others are described in § 3.

Proposition 1. *Every Fréchet space has a completing web (of closed sets).*

Proof. Let (V_n) be a base of (closed) absolutely convex neighbourhoods with $V_{n+1} \subseteq \frac{1}{2} V_n$ for each n. The web formed by taking every set in the nth layer to be V_n is clearly compatible and therefore completing, by Lemma 1. (There is essentially only one strand, (V_n) itself.)

PROPOSITION 2. *The strong dual of a metrisable convex space has a completing web (of closed sets).*

Proof. Let (U_i) be a base of neighbourhoods in the metrisable convex space, with dual F. For the first layer of the web in F, take the sequence (U_i^0) of polars; then take the sets of the second layer determined by U_i^0 all to be $\frac{1}{2}U_i^0$, and in general the sets of the nth layer to be $2^{-n+1}U_i^0$. Since each U_i^0 is strongly bounded, the web is compatible with the strong topology, which is complete (Ch. VI, Prop. 1, Cor. 1). Hence by Lemma 1, the web is completing. (In this case there are essentially countably many strands, $(2^{-n+1}U_i^0)$, one for each i.)

Readers who know something about topological groups may like to keep another interpretation in mind. To each strand (W_n) of a web corresponds a topology, ζ say, defined by taking $\{x+W_n\}$ to be a base of neighbourhoods of x for each $x \in F$. Under ζ, F is an additive topological group (not, in general, a topological vector space, since the W_n may fail to be absorbent). Saying that the web is compatible with the convex space topology η of F is the same as saying that each ζ is finer than η; in that case, under each ζ, F is a separated metrisable topological group. It is easy to see that $\mathscr{W}$ is then a completing web if and only if, for each strand, every ζ-Cauchy sequence in F is η-convergent. Further, if $\mathscr{W}$ is a completing web of closed sets (as it will usually be in specific cases) then, for each strand topology ζ, F is a ζ-complete metrisable topological group. We revert to these ideas later and in Supplement 3.

2. The closed graph theorem. The chief aim of this section is to prove a closed graph theorem for a linear mapping of an inductive limit of Banach spaces into a convex space with a completing web. The main steps are simple adaptations of the proofs of the theorems of Banach mentioned in Supplement 2 of Chapter VI. The arguments involve category; the necessary definitions are in Supplement 1 of Chapter IV. In particular, the lemma that follows depends on the fact that the union of a sequence of meagre subsets of a topological space is meagre, which is immediate from the definition.

LEMMA 2. *Let $\mathscr{W}$ be any web in a convex Baire space E. Then there is a strand (W_n) such that, for each n, $\overline{W}_n$ is a neighbourhood of the origin in E.*

Proof. We define a strand (W_n) for which each set W_n is not meagre. Since E is a Baire space, it is not meagre and so we can start with $W_0 = E$. Suppose then that W_r has been defined for all $r < n$. If B_j are the sets of the nth layer of $\mathscr{W}$ determined by W_{n-1}, the union of all the sets mB_j, as m and j vary, contains W_{n-1}, which by hypothesis is not meagre. Hence at least one B_j is not meagre; take this set to be W_n.

Now for each n, the absolutely convex set $\overline{W}_n$ contains an interior point and so (cf. p. 74) is a neighbourhood of the origin in E.

In order to apply Lemma 2 when E is a Fréchet space, or a Banach space, we need to know that these are Baire spaces. This is a consequence of Baire's category theorem, already mentioned in the Supplement on category; to make this account self-contained we outline the proof.

THEOREM 1. *Every complete metric space is a Baire space.*

Proof. Suppose, on the contrary, that the open subset A is the union of a sequence of nowhere dense subsets A_n; thus no $\overline{A}_n$ contains a ball. Then there is some closed ball B_1 contained in A but disjoint from A_1 (since otherwise every point of A would lie in $\overline{A}_1$). Next, there is a closed ball B_2 contained in B_1, but disjoint from A_2. (For if every ball in B_1 met A_2, every point of B_1 would lie in $\overline{A}_2$.) This process can be continued to construct a decreasing sequence (B_n) of closed balls, with radii converging to zero, such that each B_n is disjoint from A_n. Their centres form a Cauchy sequence, convergent to x, say; then x is in all B_n and therefore not in any A_n. Hence $x \notin A$, but $x \in B_1 \subseteq A$, a contradiction.

THEOREM 2. *Let E be a Fréchet space and F a separated convex space with a completing web. If t is a linear mapping of E into F with a closed graph then t is continuous.*

Proof. The inverse images by t of the sets of the web $\mathscr{W}$ in F form a web in E and therefore, by Lemma 2, there is a strand (W_n) of $\mathscr{W}$ such that each $\overline{t^{-1}(W_n)}$ is a neighbourhood of the origin in E. Thus there is a sequence of neighbourhoods U_n, which we may take to be a base, such that $U_n \subseteq \overline{t^{-1}(W_n)}$ for each n. Hence, for all neighbourhoods U in E, $U_n \subseteq t^{-1}(W_n) + U$ and so, for all n,

$$t(U_n) \subseteq t(U_{n+1}) + W_n.$$

Now take any closed absolutely convex neighbourhood V in F. Since the web $\mathscr{W}$ is compatible, $W_{n-1} \subseteq V$ for some n; we show that $t(U_n) \subseteq V$, which will prove t continuous. So take any $x_0 \in U_n$. By the formula just proved, there exists $x_1 \in U_{n+1}$ with $t(x_0) - t(x_1) \in W_n$, again there exists $x_2 \in U_{n+2}$ with $t(x_1) - t(x_2) \in W_{n+1}$ and so on. In general we choose $x_r \in U_{n+r}$ so that $t(x_r) - t(x_{r+1}) \in W_{n+r}$. Since $\mathscr{W}$ is completing, $\Sigma(t(x_r) - t(x_{r+1}))$ is convergent and therefore $t(x_r) \to y$, say. But $x_r \to o$, so that $y = o$, the graph of t being closed. Thus $t(x_r) \to o$. Also

$$t(x_0) - t(x_r) = \sum_{s<r} (t(x_s) - t(x_{s+1})) \in W_n + \ldots + W_{n+r-1} \subseteq W_{n-1}.$$

Hence $t(x_0) \in \overline{W_{n-1}} \subseteq \overline{V} = V$.

The above proof shows that (with the same notation) there exist a strand (W_n) and a base (U_n) of neighbourhoods such that $t(U_n) \subseteq \overline{W}_n$. In particular, if the sets of the web are closed, as is often the case, the image $t(E)$ lies in the intersection of the vector subspaces spanned by the sets of some strand. Also in this case t is continuous when F is given the group topology ζ defined by the strand (W_n). More generally, this last result continues to be true whenever F is ζ-complete (for then $\Sigma(t(x_r) - t(x_{r+1}))$ converges under ζ).

It is worth pointing out that the proof of Theorem 2 uses only the fact that the graph of t is sequentially closed, that is, that $x_n \to x$ and $t(x_n) \to y$ imply $y = t(x)$. The advantage of this slight relaxation in the hypotheses lies not so much in greater generality but in the ease of verification, especially for non-metrisable spaces.

THEOREM 3. *Suppose that E is the inductive limit of a family of Banach spaces and F is a separated convex space with a completing* *web. If t is a linear mapping of E into F with a (sequentially) closed graph, then t is continuous.*

Proof. Let E be the inductive limit of the spaces E_γ by mappings u_γ. If $x_n \to x$ in E_γ and $t(u_\gamma(x_n)) \to y$, then $u_\gamma(x_n) \to u_\gamma(x)$ by the continuity of u_γ and so $y = t(u_\gamma(x))$, since t has sequentially closed graph. Thus the graph of $t \circ u_\gamma$ is sequentially closed and therefore, by Theorem 2, $t \circ u_\gamma$ is continuous. Hence (Ch. v, Prop. 5) t is continuous, as required.

We can readily obtain an open mapping theorem from this. With the same hypotheses on E and F, let t be a continuous linear mapping of F onto E. Then t is open. To prove this, write $t = s \circ k$, where k is the canonical mapping of F onto the separated quotient $F_1 = F/t^{-1}(o)$ and s is a continuous (1,1) linear mapping of F_1 onto E. Now F_1 has a completing web (the images by k of the sets of the given completing web in F form one; see the next section) and the result now follows by applying the theorem to s^{-1}.

3. The scope of the theorems. In order to be able to apply the main results of the previous section, it is necessary to know which spaces have completing webs and which are inductive limits of Banach spaces. Some information on the first of these questions is already available in Propositions 1 and 2, which show that Fréchet spaces and their strong duals have completing webs. So also do strict inductive limits of Fréchet spaces and their strong duals, as is shown at the end of this section. First, however, we take up the question of which spaces are inductive limits of Banach spaces. It can already be seen that these include separated Mackey (bornological) spaces that are complete, or even only sequentially complete (Ch. v, Th. 1); indeed the term *ultra-bornological* is sometimes used for spaces that are inductive limits of Banach spaces. Thus Fréchet spaces and inductive limits of Banach spaces qualify (Ch. v, Prop. 8). The same holds for many (but not all) of the strong duals of such spaces; the results of this section deal with duals of reflexive Fréchet spaces and of strict inductive limits of Banach spaces or reflexive Fréchet spaces. The effect of all this is to show that a closed graph theorem is available for most of the useful spaces encountered in functional analysis (see Supplement 1).

Proposition 3. *Let E be a metrisable convex space such that the dual E' of E is barrelled under $\beta(E', E)$. Then E' is an inductive limit of Banach spaces.*

Proof. Since E' is complete (Ch. VI, Prop. 1, Cor. 1) it is sufficient to show that it is a Mackey space.

Let t be a linear mapping of E' into a convex space F. We have to show that if t is bounded (i.e. takes bounded sets into bounded sets), then t is continuous (cf. Ch. V, §3). Let (U_n) be a base of neighbourhoods in E and let V be any absolutely convex neighbourhood in F. Since each U_n^0 is bounded in $\beta(E', E)$, $t(U_n^0)$ is bounded in F and so there is some $\lambda_n > 0$ with $t(\lambda_n U_n^0) \subseteq V$. Let $A_n = \sum_{1 \leqslant r \leqslant n} 2^{-r}\lambda_r U_r^0$; then each A_n is absolutely convex and $t(A_n) \subseteq V$, so that, if $A = \bigcup A_n, t(A) \subseteq V$. The proof will therefore be complete if we show that A is a β-neighbourhood. Now each A_n is $\sigma(E', E)$-compact (Ch. III, Th. 6 and Lemma 7) and so $\beta(E', E)$-closed; also since $E' = \bigcup U_n^0$, A is absorbent. The result now follows on applying the next lemma, of interest in itself, to E' with the strong topology.

Lemma 3. *Let E be a barrelled space and (A_n) an increasing sequence of closed absolutely convex subsets whose union A is absorbent. Then A is a neighbourhood in E.*

Proof. It is sufficient to show that $\bar{A} \subseteq 2A$, since $\bar{A}$ is a barrel. If $x \notin 2A$ then for each n, $x \notin 2A_n$. Hence there exists $x_n' \in A_n^0$ such that $\langle x, x_n' \rangle = 2$. Now $B' = \{x_n' : n \geqslant 1\}$ is $\sigma(E', E)$-bounded. For if $y \in E$ then there exist some n and λ so that $y \in \lambda A_n$. Since (A_n) is increasing, (A_n^0) is decreasing and so $x_m' \in A_n^0$ for $m \geqslant n$. Also $\{x_1', \ldots, x_{n-1}'\}$ is a finite set; thus $\{\langle y, x_m' \rangle : m \geqslant 1\}$ is bounded. Since E is barrelled, B' is equicontinuous and so $\bar{B}'$ is $\sigma(E', E)$-compact; hence (x_n') has a cluster point, x' say. Then, since $x_m' \in A_n^0$ for $m \geqslant n$, $x' \in A_n^0$ for all n and so $x' \in A^0$; also $\langle x, x' \rangle = 2$, so that $x \notin \bar{A}$.

Corollary (of Proposition 3). *The strong dual of a reflexive Fréchet space is an inductive limit of Banach spaces.*

For it is barrelled (Ch. IV, Prop. 4).

Proposition 4. *Let $E = \bigcup E_n$ be the strict inductive limit of a sequence of metrisable spaces, such that for each n, the strong dual E_n' is barrelled. Then E' is an inductive limit of Banach spaces.*

Proof. Again, E' is complete (Ch. VI, Prop. 1, Cor. 1; Ch. V, Props. 7 and 8); we show that E' is a Mackey space.

Let t be a bounded linear mapping of E' into a convex space F, and let V be any absolutely convex neighbourhood in F, with gauge q. Then $N = q^{-1}(0)$ is a closed vector subspace of F contained in V; let k be the canonical mapping of F onto F/N and let $v = k \circ t$.

First of all, there is an integer m for which $v(E_m^0) = \{o\}$. For if not, there are points $z_n' \in E_n^0$ for each n with $t(z_n') \notin nV$. Since the restrictions of all but a finite number of the z_r' to E_n are zero, $\{z_n'\}$ is equicontinuous (Ch. V, Prop. 5) and so strongly bounded. But this implies that $\{t(z_n')\}$ is bounded, a contradiction.

Next, any absolutely convex neighbourhood U_m in E_m is of the form $U \cap E_m$ for some absolutely convex neighbourhood U in E (Ch. VII, Prop. 1). By the Hahn–Banach theorem (Ch. II, Th. 3) each $x_m' \in U_m^0$ is the restriction to E_m of at least one $x' \in U^0$. Moreover, by the first part of the proof, if x' and y' have the same restriction to E_m, then $v(x') = v(y')$. We may therefore define a linear mapping v_m of E_m' to F/N by putting $v_m(x_m') = v(x')$ for any (and all) x' which give x_m' on restriction to E_m. Further, if B is a bounded subset of E_m', B is equicontinuous (Ch. V, Props. 8 and 9) and therefore contained in some $(U \cap E_m)^0$; thus

$$v_m(B) \subseteq v(U^0) = k(t(U^0)),$$

a bounded set. This proves that v_m is bounded.

Now Proposition 3 shows that v_m is continuous and so there is a bounded subset A of E_m with $v_m(A^0) \subseteq k(V)$. Hence

$$t(A^0) \subseteq V + k^{-1}(o) \subseteq 2V,$$

which completes the proof that t is continuous.

COROLLARY. *The strong dual of a strict inductive limit of a sequence of Banach spaces or of reflexive Fréchet spaces is an inductive limit of Banach spaces.*

We turn now to examine the stability properties of the class of spaces with completing webs.

First, let F be a separated space with a completing web $\mathscr{W}$. Then if M is a closed vector subspace of F, a completing web in M is formed by the sets $M \cap A$ as A runs through $\mathscr{W}$. Also any separated continuous image of F has a completing web, consisting of the images of the sets of $\mathscr{W}$. Thus in particular any separated quotient of F has a completing web.

Next, let (F_n) be a sequence of spaces with completing webs. Then the inductive limit $F = \bigcup v_n(F_n)$ of the spaces F_n by the mappings v_n, supposed separated, has a completing web $\mathscr{W}$. For each $v_n(F_n)$, under the topology induced on it by that of F, has a completing web $\mathscr{W}_n$ and now $\mathscr{W}$ is formed by laying the $\mathscr{W}_n$ side by side, so that the rth layer of $\mathscr{W}$ is composed of all the rth layers of the $\mathscr{W}_n$, and each strand of $\mathscr{W}$ is a strand of some $\mathscr{W}_n$.

Still supposing that each F_n has a completing web, we see readily that any finite product $F_1 \times \ldots \times F_n$ has a completing web, the sets in the rth layer being the products of sets in the rth layers from $F_1, \ldots, F_n$. The direct sum ΣF_n has a completing web, since it is the inductive limit of a sequence of such finite products.

In each of the above cases the verification of the absorbency and convergence conditions presents no difficulties.

To form a web for an infinite product this construction has to be modified, so that the factor spaces are brought in one by one at successive layers. This is a special case of a projective limit; so also is the strong dual of a strict inductive limit of a sequence of Fréchet spaces (Ch. v, Prop. 15 and Ch. vii, Prop. 4).

PROPOSITION 5. *Let F be the projective limit by mappings v_n of a sequence of separated convex spaces F_n. If each F_n has a compatible web $\mathscr{W}_n$, then F has a compatible web $\mathscr{W}$. Suppose also that*

(*) *whenever (x_r) is a sequence in F such that $(v_n(x_r))$ is convergent in F_n for each n, then (x_r) is convergent in F.*

Then if each $\mathscr{W}_n$ is completing, so is $\mathscr{W}$.

Proof. For the first layer of $\mathscr{W}$, take the sets $v_1^{-1}(A_i^{(1)})$, where $(A_i^{(1)})$ is the first layer in $\mathscr{W}_1$. For the sets of the second layer determined by $v_1^{-1}(A_i^{(1)})$, take the sets $v_1^{-1}(A_{ij}^{(1)}) \cap v_2^{-1}(A_r^{(2)})$, $(j, r = 1, 2, \ldots)$, where $(A_r^{(2)})$ is the first layer of $\mathscr{W}_2$. Continue in this way, so that a strand (W_n) is given by

$$W_n = v_1^{-1}(W_n^{(1)}) \cap v_2^{-1}(W_{n-1}^{(2)}) \cap v_3^{-1}(W_{n-2}^{(3)}) \cap \ldots \cap v_n^{-1}(W_1^{(n)}),$$

where $(W_n^{(i)})$ is a strand in $\mathscr{W}_i$ for each i. Now if V is a neighbourhood in F, then $V = \bigcap_{1 \leqslant i \leqslant n} v_i^{-1}(V_i)$ for some n, where v_i is a neighbourhood in F_i. Since each $\mathscr{W}_i$ is compatible, there is some $m \geqslant n$ so large that

$$W_m^{(1)} \subseteq V_1,\ W_{m-1}^{(2)} \subseteq V_2,\ \ldots,\ W_{m-n+1}^{(n)} \subseteq V_n.$$

Then $W_m \subseteq V$ and $\mathscr{W}$ is compatible.

Suppose that (*) holds and that $y_n \in W_n$ for each n. Then, for each i, $\sum_{n \geqslant i} v_i(y_n)$ converges, since $\mathscr{W}_i$ is completing. Writing $x_r = \sum_{1 \leqslant n \leqslant r} y_n$, we have $(v_i(x_r))$ convergent in F_i for each i. Hence, by (*), Σy_n converges, and $\mathscr{W}$ is completing.

Corollary 1. *If F is sequentially complete and is the projective limit of a sequence of spaces with compatible webs then F has a completing web.* (By Lemma 1.)

Corollary 2. *The product $F = \mathsf{X} F_n$ of a sequence of spaces with completing webs has a completing web.*

For then v_n is the projection p_n of F onto F_n; if $p_n(x_r) \to y_n$, say, for each n, then $x_r \to x = (y_n)$, so that (*) holds.

Corollary 3. *Suppose that (F_n) is a decreasing sequence of separated convex spaces with completing webs, such that the topology on each F_{n+1} is finer than that induced on it by F_n, and let $F = \bigcap F_n$ have the projective limit topology, the coarsest finer than those induced by all the F_n. Then F has a completing web.*

Proof. We show that (*) is satisfied with v_n the injection of F into F_n. If (x_r) is any sequence in F, convergent in each F_n, to y_n say, then clearly $y_n = y_{n+1}$ for all n and so

$$y_1 = y_2 = \ldots = y_n = \ldots = y \in F.$$

Now $x_r \to y$ in each F_n and so $x_r \to y$ in the projective limit topology on F.

Propositions 1 and 2 show that Fréchet spaces and their strong duals have completing webs. From what we have just proved, it follows that so do a strict inductive limit of a sequence of Fréchet spaces and its strong dual (Prop. 3, Cor.). Finally, we mention without proof that, more generally, a space $L(E, F)$ of continuous linear mappings, under any topology of uniform convergence, may be shown to have a completing web of closed sets, when E is a Fréchet space or a strict inductive limit of a sequence of Fréchet spaces, and F has a completing web of closed sets.

SUPPLEMENT

(1) *Specific spaces.* Most of the spaces of the Supplements to Chapters I and II are inductive limits of Banach spaces and also possess completing webs. In particular, the space $\mathscr{D}$ of indefinitely differentiable functions is the strict inductive limit of a sequence of reflexive Fréchet spaces (Ch. VII, Suppl. 1); therefore, by the above results, $\mathscr{D}$ and its strong dual $\mathscr{D}'$, the space of distributions, have both these properties. So also do the Fréchet spaces $\mathscr{E}$, $\mathscr{S}$ and $\mathscr{H}(D)$, and their strong duals. The space $\mathscr{K}(]-\infty,\infty[)$ is a strict inductive limit of a sequence of Banach spaces; thus its strong dual is also an inductive limit of Banach spaces, and both the space and its dual possess completing webs.

(2) *Open mapping theorem.* The simplest form of this theorem, given after Theorem 3, asserts that, under certain conditions on E and F, a linear mapping t of F onto E is open if it is continuous. The same method shows that, more generally, t is open if its graph is closed (cf. Ch. VI, Prop. 10 (ii)). Indeed, it is enough to assume the graph sequentially closed. (First, an argument similar to that of Theorem 2 shows that if t maps a vector subspace F_0 of a space F with a completing web onto a Fréchet space E, and if the graph of t is sequentially closed (in $F \times E$) then t is open. Now, with the hypotheses on E and F of Theorem 3, suppose that the graph of t is sequentially closed. Since the continuous open image of a Banach space is a Banach space (Ch. VI, Prop. 13), each E_γ may be taken to be a vector subspace of E (Ch. V, end of p. 80). Then the restriction t_γ of t to $F_\gamma = t^{-1}(E_\gamma)$ maps F_γ onto the Banach space E_γ and it is easy to see that the graph of t_γ is sequentially closed in $F \times E_\gamma$. By the above theorem, each t_γ is open. Since $F = \bigcup F_\gamma$ and E is the inductive limit of the spaces E_γ, it follows that t is open.)

(3) *Space of subsets.* The closed graph theorems in Chapter VI and those in this Appendix are closely connected; they can be approached in the same way by the use of uniform structures on

the space of subsets of the range space F (see Ch. VI, Suppl. 3). We sketch this method, taking up again the use of the group topology associated with each strand of a web.

Suppose that $\mathscr{W}$ is a filter base of absolutely convex subsets of a vector space F. There corresponds a uniform structure, $\dot{\omega}$ say, on the space of all non-empty subsets of F, defined by the vicinities

$$\dot{W} = \{(A, B) \colon A \subseteq B + W \quad \text{and} \quad B \subseteq A + W\}$$

for each $W \in \mathscr{W}$. If $\mathscr{W}$ is a base of absolutely convex neighbourhoods in a locally convex topology η on F, we get the Hausdorff uniform structure, which we denote by $\dot{\eta}$. If $\mathscr{W}$ is one strand of a web in F, we get the Hausdorff uniform structure $\dot{\zeta}$ defined by the associated group topology ζ. In either case, if F is also complete and metrisable, this uniform structure coincides on the space of closed subsets with that derived from the Hausdorff distance. The space of closed subsets, and so also the (non-separated) space of all subsets, is then complete. (See, for example, Bourbaki, *Topologie générale*, Ch. IX, §2, exer. 7.)

Suppose now that E is a convex space with a base $\mathscr{U}$ of absolutely convex neighbourhoods, that F is a separated convex space with topology η and a base $\mathscr{V}$ of absolutely convex neighbourhoods, and that $\mathscr{W}$ is a filter base of absolutely convex subsets of F. Let t be a linear mapping of E into F. Then t is continuous if (and only if) the net $t(\mathscr{U})$ in the space of subsets of F, directed by inclusion in F, is $\dot{\eta}$-convergent to $\{o\}$. It is easy to show that if $t(\mathscr{U})$ is $\dot{\eta}$-convergent, then it converges to

$$\bigcap\{\overline{t(U)} \colon U \in \mathscr{U}\}.$$

Also $\bigcap \overline{t(U)} = \{o\}$ if (and only if) the graph of t is closed (cf. Ch. VI, proof of Lemma 4). Now $t(\mathscr{U})$ is $\dot{\omega}$-Cauchy if (and only if) $\overline{t^{-1}(W)}$ is a neighbourhood in E for each $W \in \mathscr{W}$ (as in Theorem 2). Hence a closed graph theorem will hold if, in E, $\overline{t^{-1}(W)}$ is a neighbourhood for each $W \in \mathscr{W}$, and, in F, the fact that $t(\mathscr{U})$ is $\dot{\omega}$-Cauchy implies that it is $\dot{\eta}$-convergent.

When E is barrelled and $\mathscr{W}$ is a base of neighbourhoods in η, each $W \in \mathscr{W}$ is absorbent and so $\overline{t^{-1}(W)}$ is a barrel, therefore a neighbourhood in E. When E is a Baire space then, by Lemma 2, there is a strand (W_n) such that each $\overline{t^{-1}(W_n)}$ is a neighbourhood in E.

Corresponding to these two cases, suppose first that $\mathscr{W}$ is a base of neighbourhoods in η on F, so that $\dot{\omega} = \dot{\eta}$. If F is hypercomplete (Ch. VI, Suppl. 3) then, each $t(U)$ being absolutely convex, $t(\mathscr{U})$ is $\dot{\eta}$-convergent if it is $\dot{\eta}$-Cauchy. This may also be shown true if F is fully complete. Thus we reach Theorem 6 of Chapter VI. Next, suppose that $\mathscr{W}$ is a strand of a completing web. If the sets of the web are closed, or if, more generally, F is ζ-complete, then the space of subsets of F is $\dot{\zeta}$-complete; in particular, if $t(\mathscr{U})$ is $\dot{\zeta}$-Cauchy then it is $\dot{\zeta}$-convergent and so $\dot{\eta}$-convergent. We mentioned earlier that F has a completing web if and only if, for each strand, every ζ-Cauchy sequence in F is η-convergent; it can be shown that this is equivalent to the $\dot{\eta}$-convergence of every $\dot{\zeta}$-Cauchy sequence, or net, of (absolutely convex) subsets of F. In the closed graph theorem that results here, E is not assumed metrisable, as it is in Theorem 2; on the other hand, we make use of the fact that the graph is closed, rather than merely sequentially closed. (For a detailed approach along these lines, see *Proc. London Math. Soc.* (3), 24 (1972) 692–738.)

(4) *Further generalisations.* With appropriate minor modifications, the results of this Appendix remain valid for non-locally convex spaces, provided that the definition of a web is suitably adjusted. Instead of demanding absolutely convex sets with $W_{n+1} \subseteq \frac{1}{2}W_n$, we simply require $W_{n+1} + W_{n+1} \subseteq W_n$ for all strands. In locally convex spaces too, the hypothesis that the sets of a web must be absolutely convex may be dropped, provided that the absorbency and convergence conditions are altered; the effect is to make a web resemble a sifting on a Souslin space. By generalisations such as these, but often at the expense of greater complication, the closed graph theorem may be proved for a wider class of range spaces.

Further developments are to be found in the Mémoire by M. De Wilde cited at the beginning of this Appendix.

BIBLIOGRAPHY

There are now plenty of books from which information can be obtained about the theory of topological vector spaces and its applications to the theory of distributions. Some of these are listed here, together with a few books on functional analysis. Banach's book **1** is a classic; **3**, **4**, **12** and **15** are general books on functional analysis and all but **12** contain sections on locally convex spaces. For the theory of distributions, see **14**. Of the books on topological vector spaces, **5**, **6**, **8** and **16** emphasise the connection with distributions and **2**, **7**, **10**, **11** and **13** are general accounts of the theory. Finally, **8** treats a specialised topic.

1 Banach, S. *Théorie des opérations linéaires* (Warsaw, 1932).

2 Bourbaki, N. *Éléments de mathématique*. Livre V. *Espaces vectoriels topologiques* (Paris, Hermann, 1953, 1955).

3 Day, M. M. *Normed linear spaces* (Berlin, Springer-Verlag, 1958).

4 Dunford, N. and Schwartz, J. *Linear operators*. Vol. I. *General theory* (New York, Interscience, 1958).

5 Edwards, R. E. *Functional analysis, theory and applications* (New York, Holt, Rinehart and Winston, 1965).

6 Garsoux, J. *Espaces vectoriels topologiques et distributions* (Paris, Dunod, 1963).

7 Grothendieck, A. *Espaces vectoriels topologiques* (São Paulo, 1954).

8 Grothendieck, A. *Produits tensoriels topologiques et espaces nucléaires* (Mem. Amer. Math. Soc. no. 16, 1955).

9 Horváth, J. *Topological vector spaces and distributions*, Vol. I (Reading, Addison-Wesley, 1966).

10 Kelley, J. L., Namioka, I. and co-authors. *Linear topological spaces* (Princeton, Van Nostrand, 1963).

11 Köthe, G. *Topological vector spaces* (trans. D. J. H. Garling) (Berlin, Springer-Verlag, 1969).

12 Riesz, F. and Sz.-Nagy, B. *Functional analysis* (trans. L. Boron) (London, Blackie, 1956).

13 Schaefer, H. H. *Topological vector spaces* (New York, Macmillan, 1966).

14 Schwartz, L. *Théorie des distributions* (Paris, Hermann, 1950, 1951).

15 Taylor, A. E. *Introduction to functional analysis* (New York, Wiley, 1958).

16 Trèves, F. *Topological vector spaces, distributions and kernels* (New York, Academic Press, 1967).

Many of the above books contain extensive bibliographies (e.g. **3**, **4**, **5**, **9**, **11** and **13**) and so we have attempted no more than to list below some of the papers that were instrumental in developing the subject.

ARENS, R. Duality in linear spaces. *Duke Math. J.* 14 (1947), 787–94.

BOURBAKI, N. Sur certains espaces vectoriels topologiques. *Ann. Inst. Fourier Grenoble*, 2 (1950), 5–16 (1951).

DIEUDONNÉ, J. La dualité dans les espaces vectoriels topologiques. *Ann. Sci. École Norm. Sup.* (3), 59 (1942), 107–39.

DIEUDONNÉ, J. Recent developments in the theory of locally convex vector spaces. *Bull. Amer. Math. Soc.* 59 (1953), 495–512.

DIEUDONNÉ, J. and SCHWARTZ, L. La dualité dans les espaces ($\mathscr{F}$) et ($\mathscr{L}\mathscr{F}$). *Ann. Inst. Fourier Grenoble*, 1 (1949), 61–101 (1950).

GROTHENDIECK, A. Résumé des resultats essentiels dans la théorie des produits tensoriels topologiques et des espaces nucléaires. *Ann. Inst. Fourier Grenoble*, 4 (1952), 73–112 (1954).

GROTHENDIECK, A. Sur les espaces (F) et (DF). *Summa Brasil. Math.* 3 (1954), 57–123.

KOLMOGOROFF, A. Zur Normierbarkeit eines allgemeinen topologischen linearen Raumes. *Studia Math.* 5 (1934), 29–33 (1935).

KÖTHE, G. Die Stufenräume, eine einfache Klasse linearer vollkommener Räume. *Math. Z.* 51 (1948), 317–45.

KÖTHE, G. Neubegründung der Theorie der vollkommenen Räume. *Math. Nachr.* 4 (1951), 70–80.

MACKEY, G. W. On infinite-dimensional linear spaces. *Trans. Amer. Math. Soc.* 57 (1945), 155–207.

MACKEY, G. W. On convex topological linear spaces. *Trans. Amer. Math. Soc.* 60 (1946), 519–37.

NEUMANN, J. VON. On complete topological spaces. *Trans. Amer. Math. Soc.* 37 (1935), 1–20.

WEHAUSEN, J. V. Transformations in linear topological spaces. *Duke Math. J.* 4 (1938), 157–69.

INDEX